PRACTICAL PROBLEMS in MATHEMATICS
for DRAFTING AND CAD
2nd Edition

Delmar's *PRACTICAL PROBLEMS in MATHEMATICS* Series

- *Practical Problems in Mathematics for Automotive Technicians, 4e*
George Moore
Order # 0-8273-4622-0

- *Practical Problems in Mathematics for Carpenters, 6e*
Harry C. Huth
Order # 0-8273-4579-8

- *Practical Problems in Mathematics for Drafting and CAD, 2e*
John C. Larkin
Order # 0-8273-1670-4

- *Practical Problems in Mathematics for Electricians, 5e*
Herman and Garrard
Order 0-8273-6708-2

- *Practical Problems in Mathematics for Electronic Technicians, 3e*
Herman and Sullivan
Order # 0-8273-6761-9

- *Practical Problems in Mathematics for Graphic Artists*
Vermeersch and Southwick
Order # 0-8273-2100-7

- *Practical Problems in Mathematics for Health Occupations*
Louise Simmers
Order # 0-8273-6771-6

- *Practical Problems in Mathematics for Heating and Cooling Technicians, 2e*
Russell B. DeVore
Order # 0-8273-4062-1

- *Practical Problems in Mathematics for Industrial Technology*
Donna Boatwright
Order # 0-8273-6974-3

- *Practical Problems in Mathematics for Manufacturing, 4e*
Dennis D. Davis
Order # 0-8273-6710-4

- *Practical Problems in Mathematics for Masons, 2e*
John E. Ball
Order # 0-8273-1283-0

- *Practical Problems in Mathematics for Welders, 4e*
Schell and Matlock
Order # 0-8273-6706-6

Related Titles

- *Fundamental Mathematics for Health Careers, 3e*
Hayden and Davis
Order # 0-8273-6688-4

- *Mathematics for Plumbers and Pipefitters, 5e*
Smith, D'Arcangelo, D'Arcangelo, Guest
Order 0-7061-x

- *Vocational-Technical Mathematics, 3e*
Robert D. Smith
Order # 0-8273-6806-9

PRACTICAL PROBLEMS in MATHEMATICS
for DRAFTING AND CAD
2nd Edition

Dr. John C. Larkin

**Central Connecticut
State University**

Delmar Publishers Inc.™

I(T)P An International Thomson Publishing Company

Albany • Bonn • Boston • Cincinnati • Detroit • London • Madrid • Melbourne
Mexico City • New York • Pacific Grove • Paris • San Francisco • Singapore • Tokyo
Toronto • Washington

NOTICE TO THE READER

Cover Design: Dartmouth Publishing

Delmar Staff:
Publisher: Robert D. Lynch
Editor: Mary Clyne
Assistant Editor: Mona Kulkarni
Production Manager: Larry Main
Art & Design Coordinator: Nicole Reamer

COPYRIGHT © 1996
By Delmar Publishers Inc.
an International Thomson Publishing Company
The ITP logo is a trademark under license.

Printed in the United States of America

For more information, contact:

Delmar Publishers
3 Columbia Circle, Box 15015
Albany, New York 12212-5015

International Thomson Editores
Campos Eliseos 385, Piso 7
Col Polanco
11560 Mexico D F Mexico

International Thomson Publishing Europe
Berkshire House 168 - 173
High Holborn
London, WC1V 7AA
England

International Thomson Publishing GmbH
Königswinterer Strasse 418
53227 Bonn
Germany

Thomas Nelson Australia
102 Dodds Street
South Melbourne, 3205
Victoria, Australia

International Thomson Publishing Asia
221 Henderson Road
#05 - 10 Henderson Building
Sinapore 0315

Nelson Canada
1120 Birchmount Road
Scarborough, Ontario
Canada, M1K 5G4

International Thomson Publishing - Japan
Hirakawacho Kyowa Building, 3F
2-2-1 Hirakawacho
Chiyoda-ku, Tokyo 102
Japan

1 2 3 4 5 6 7 8 9 10 XXX 01 00 99 98 97 96 95
Library of Congress Cataloging-in-Publication Data

Larkin, John C.
    Practical problems in mathematics for drafting and CAD
    John C. Larkin — 2nd ed.
        p.   cm.
        ISBN: 0-8273-4624-7
        1. Mechanical drawing—Mathematics—Problems, exercises, etc. 2. Computer-aided design—Mathematics—
    Problems, exercises, etc. I. Title.
T354.L37   1995
604.2' 43—dc20
                                                                                95-22533
                                                                                CIP

Contents

SECTION 5 MEASUREMENT

SECTION 6 RATIO AND PROPORTION

SECTION 7 APPLIED ALGEBRA

SECTION 8 GRAPHS

SECTION 9 APPLIED GEOMETRY

SECTION 10 APPLIED TRIGONOMETRY

APPENDIX / 276

GLOSSARY / 284

ANSWERS TO ODD-NUMBERED PROBLEMS / 287

Preface

The drafting area has recently undergone dramatic changes. The single most significant change to occur is the widespread use of computer-aided drafting and design (CAD) in all the drafting fields. Drafters have been able to increase their production using the computer. CAD systems increase the speed, accuracy, quality, and repeatability of drafters. Students of drafting need to be familiar with specialized drafting tools as well as CAD software. They still must decide what information is needed to communicate the message in the form of a drawing or a computer plot.

Practical Problems in Mathematics for Drafting and CAD has been prepared to provide practical problem solving experiences and realistic mathematical problems encountered in several drafting fields. This second edition has been improved by adding concise mathematical explanations and solved problems at the beginning of each unit. This will allow the student another opportunity to review the concepts employed to solve mathematical problems. An explanation of the use of a scientific calculator is provided; however, its use is optional as determined by the instructor.

The workbook is an excellent supplement to any vocational mathematics text where realistic problems encountered in several drafting areas can be found. Drafting students will find many relevant experiences provided to test their problem solving capabilities as well as reinforcing the graphic language used in various drafting fields. In solving the problems mathematically, the student must apply the techniques of interpreting blueprints, sketches, and multiview drawings and understand the role of the CAD drafter or operator in producing drawings. The problems were designed to be practical, realistic, and challenging to students at all levels. Standardized procedures and conventional practices used in the various fields of drafting were also adhered to in the examples presented.

Practical Problems in Mathematics for Drafting and CAD is one of a series of workbooks designed to offer students practical problem-solving experience in various occupations. The workbooks offer a step-by-step approach to the mastery of basic skills of mathematics. Each workbook includes relevant and easily understood problems in a specific vocational field. The workbooks are suitable for any student from the junior high school level through high school and up to the two-year college level. A glossary is included to aid students with technical terms and an appendix provides information on measurement, formulas, and trigonometric functions. Answers to odd-numbered problems are also provided.

The Instructor's Guide provides solutions and answers to all problems found in the workbook as well as instructional aids that will be helpful to the teacher. Instructional aids include achievement reviews and a diagnostic reading survey.

The author is an experienced drafting and CAD instructor at Central Connecticut State University. He is a professor of Technology Education and is active in numerous professional organizations. Dr. Larkin earned his doctorate from the University of Maryland.

Delmar Publishers' Online Services
To access Delmar on the World Wide Web, point your browser to:
http://www.delmar.com/delmar.html
To access through Gopher: gopher://gopher.delmar.com
(Delmar Online is part of "thomson.com", an Internet site with information on more than 30 publishers of the International Thomson Publishing organization.)
For more information on our products and services:
email: info@delmar.com
Or call 800-347-7707

To the Student

Drafting is concerned with drawings of objects. Many drafters still work "on the board" using standard drafting instruments while others use a computer and CAD software to prepare working drawings. The computer is reshaping most drafting and design applications. By the year 2000, the majority of drafters and designers will use a computer rather than traditional drafting instruments to generate their drawings. These drawings must be accurately made with standardized labeling procedures. Others must be able to read and interpret the work of a drafter or CAD operator. These are some basic facts that will help you to use this text correctly.

LINES

Dimension lines show the linear distance between two points. Arrowheads at the ends of dimension lines show the end of the intended measurement. These may be on the outside, if the measurement is very small. Extension lines are light, thin lines to indicate the end of the distance to be measured. Leaders direct attention to a dimension. Center lines are light, thin broken lines of long and short dashes. These indicate the center of a circle, or part of a circle, or that an object is symmetrical. Hidden edges are shown by a dashed line.

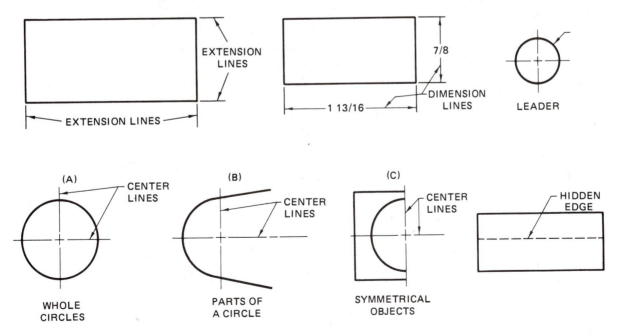

CENTER LINES

CIRCLES   The arrowhead at the end of the leader stops at the outer surface of the circle. Dimensions are given for the radius (R) or the diameter (D) of the circle or arc. The diameter is twice the radius. The circle represents a hole all the way through the object unless the depth of the hole is specified. Where holes are equally spaced on a circle, the exact location of the first hole is shown. The diameter, the size of the holes, the number of holes, and a notation, "EQUALLY SPACED," are indicated.

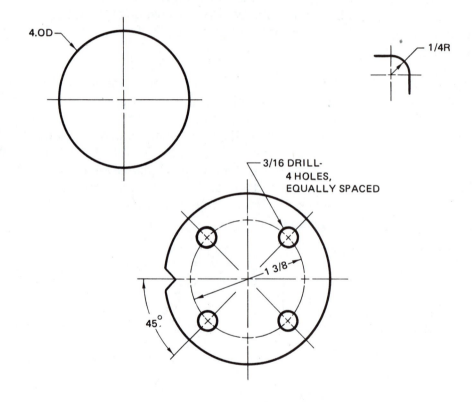

UNITS Measurement units are not used with the dimensions on the drawing. A special area is set aside for information about the object, the units of measurement, and other general information. Units are always the same within any one drawing. They are either all metric or all English.

ABBREVIATIONS

D or DIA	diameter
R or RAD	radius
TYP	typical
PLCS	places
BC	bolt circle

One drawing contains many lines:

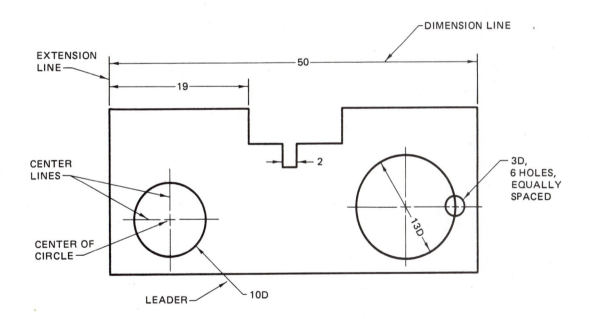

Whole Numbers

Unit 1 ADDITION OF WHOLE NUMBERS

BASIC PRINCIPLES OF ADDITION OF WHOLE NUMBERS

Whole numbers refer to complete units with no fractional parts. Addition is the process of finding the *sum* of two or more numbers. Whole numbers are added by placing them in a column with the numbers aligned on the right side of the column. The right column of numbers is added first. Write the last digit of the sum in the answer. The remaining digit(s) is carried to the next column and added. This process is followed until all columns have been added.

Example: Find the sum of 19 + 7 + 76 + 113 + 258.

```
     3              13              13
    19              19              19
     7               7               7
    76              76              76
   113             113             113
 + 258           + 258           + 258
     3              73             473
```

  19 (+) 7 (+) 76 (+) 113 (+) 258 (=) 473

PRACTICAL PROBLEMS

Add the following quantities.

1. 31 inches	2. 315 feet	3. 305 millimeters
67 inches	97 feet	117 millimeters
46 inches	216 feet	236 millimeters
+ 83 inches	+ 451 feet	+ 133 millimeters

4.   What is the overall length, in inches, of this link? _____

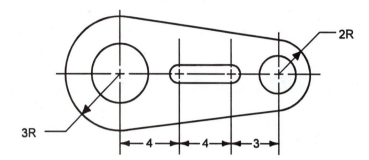

5.   A civil drafter uses a drafter's scale to measure these lengths: 6 inches, 17 inches, 34 inches, 63 inches, 26 inches, and 9 inches. What is the total length of the line? _____

6.   What is the overall length, in feet, of dimension **A** on the foundation plan? _____

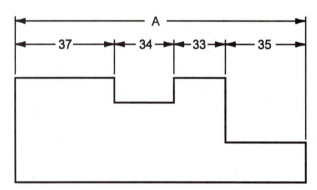

7.   A drafting department supervisor orders several types of drafting pencils for the department. The order is for 24 HB pencils, 16 H pencils, 27 2H pencils, 31 3H pencils, 48 4H pencils, 36 F pencils, 30 2B pencils, and 9 6H pencils. What is the total number of pencils? _____

8. In order to determine the drafting team's use of time, a supervisor makes this chart.

TIME RECORD

(in hours)

DRAFTER	PROJECT					HOURS (a)
	A	B	C	D	E	
Frank	6	0	3	4	1	
Jeff	3	2	7	0	5	
Bob	1	4	2	6	0	
Rod	0	6	4	2	5	
Jim	5	1	2	4	0	
TOTAL HOURS (b)						

a. Find the number of hours each drafter works.

b. Find the total number of hours the team works.

9. *Perimeter* is the total distance around the outside of a figure. What is the perimeter, in millimeters, of this shim? _____

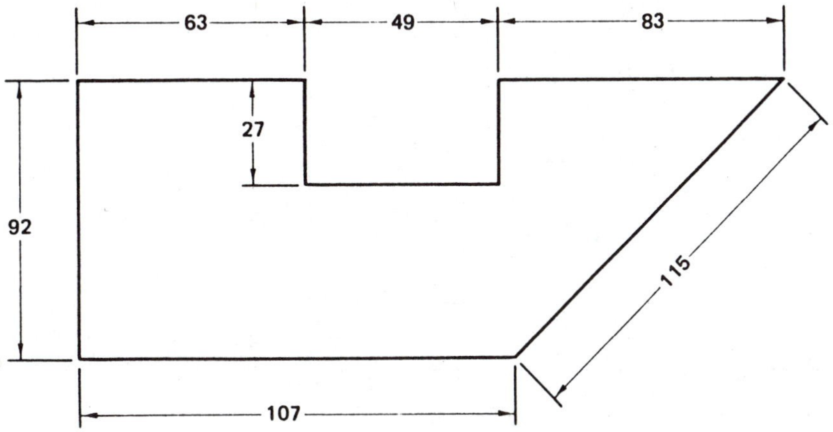

10. An office clerk must report the total number of blueprints made each week. On Monday 26 were made, 31 on Tuesday, 21 on Wednesday, 47 on Thursday, and 17 on Friday. How many blueprints were made this week?

_____

11. a. What is the length, in millimeters, of this support block?

_____

   b. What is the height, in millimeters, of this support block?

_____

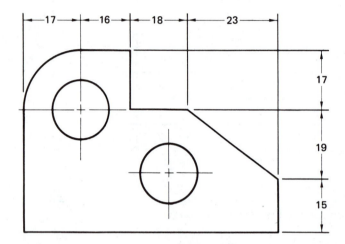

12. A map drafter is taking inventory of drafting equipment. There are 21 pencils, 6 erasers, 3 T squares, 5 triangles, 2 protractors, 7 inking pens, and 4 irregular curves. How many pieces of equipment are on hand?

_____

13. What is the perimeter, in feet, of this plate?

_____

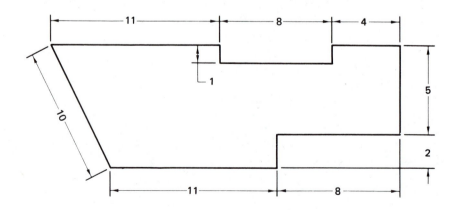

14. An architectural drafter uses many different dimensions to block in the views of an object. If the measurements are 2', 7', 11', 3', 5', and 8', find the overall length in feet.

_____

15. Find dimensions **A** and **B**, in inches.

A _____

B _____

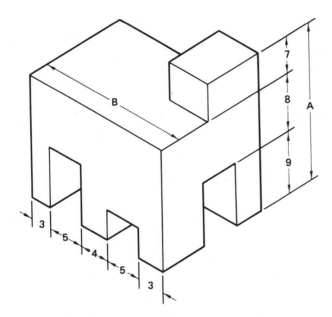

16. Find dimensions **A** and **B**, in millimeters, on this gasket.

A _____

B _____

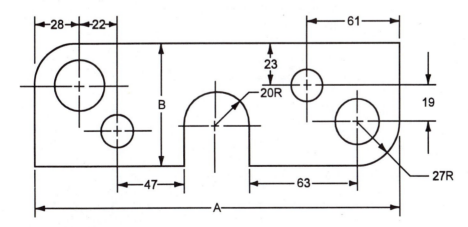

17. A detail drafter turns in the time card for five different jobs. It shows how long, in minutes, each job takes. The times, in minutes, are 236, 757, 418, 357, and 132. What is the total time worked? Express your answer in minutes.

_____

18. An illustration of a shaft is shown. Find the lengths, in millimeters, of **A**, **B**, **C**, and **D**.

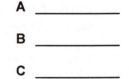

A _____

B _____

C _____

D _____

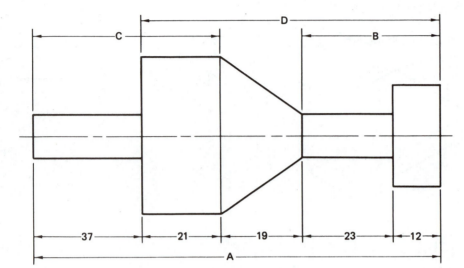

19. An assembly drawing contains six sheets of detail drawings. The number of details for these sheets are 6, 11, 9, 3, 8, and 7. What is the total number of details?

_____

20. Find the total length (dimension **G**), in millimeters, of this gauge.

_____

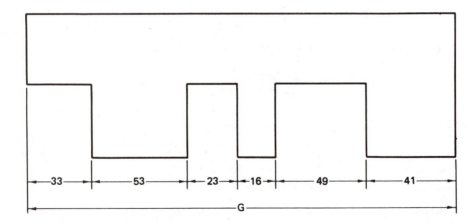

21. A computer keyboard contains 26 letters, 10 numerals, 12 function keys, 9 numeric pad keys, 11 punctuation keys, and 28 miscellaneous keys. What is the total number of selections represented by the keyboard keys mentioned above?

_____

22. A CAD input tablet menu contains 24 crosshatching symbols, 36 architectural symbols, 18 dimensioning commands, 15 editing commands, and 9 inquiry commands. What is the sum of the commands and symbols contained under these five categories on the tablet menu? _____

23. An architectural drafter needs to calculate the number of doors of each type for the door schedule for a new project. The drawings contain the following door types: 47 sliding, 35 accordion, 17 bi-fold, 15 french, 23 dutch, and 11 pocket doors. What is the sum of these doors? _____

24. A CAD drafter obtains a directory of drawing files on a floppy diskette. The files contain the following number of bytes: 239,675; 81,339; 347,021; 5,953; 67,307; 176,354, and 1,739 bytes.

    a. Find the sum of the three smallest files. _____

    b. Find the sum of the three largest files. _____

    c. Find the total number of bytes for all drawing files. _____

25. Using the absolute coordinate system presented in the illustration, provide the coordinates for each point inside the parentheses as shown. Also find (a) the height of the larger figure; (b) the width of the larger figure; (c) the height of the smaller figure; and (d) the width of the smaller figure.

a. _____

b. _____

c. _____

d. _____

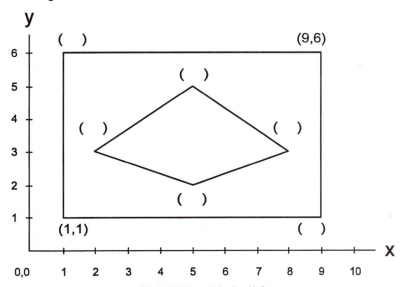

Reprinted with permission from McGrew,
Exploring the Power of AutoCAD, copyright 1990 by Delmar Publishers.

26. Find the perimeter of the CAD drawing presented if the unit space (spacing of grid points) is three inches. _____

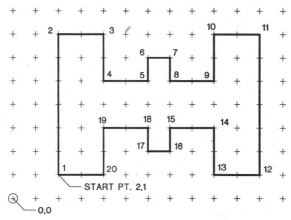

Reprinted with permission from Stellman, Krishnan, and Rhea, *Harnessing AutoCAD*, copyright 1993 by Delmar Publishers

27. a. Determine the perimeter in feet for the CAD drawing given. _____

    b. Find the sum of the three shortest boundary lines. _____

    c. Find the sum of the four longest boundary lines. _____

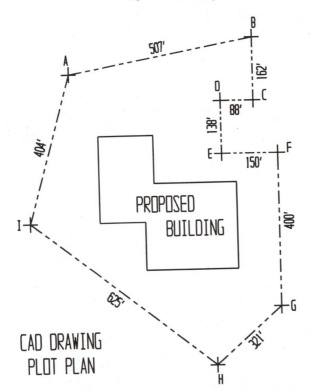

CAD DRAWING
PLOT PLAN

28. a. Determine the perimeter of the CAD drawing shown. _____

b. Find the sum of the perimeters of the five internal features. The gasket is a symmetrical part. _____

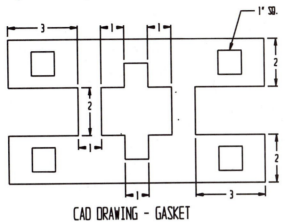

CAD DRAWING - GASKET

29. Using the CAD drawing illustrated, find the overall length and height of the house shown in elevation. Include the roof overhang.

Height _____

Width _____

CAD DRAWING

30. A CAD drafter obtains a directory of all part and pattern files used with a
     new project. Determine how many bytes are contained in part (PRT) and
     pattern (PTN) files. The directory of files appears as follows:

BASE.PRT	46,971	HEXNUT.PTN	1,124
BLOCK.PRT	31,921	LEVER.PRT	3,112
CAPSCREW.PTN	2,874	PUNCH.PRT	5,362
CAPSCR1.PTN	2,372	SHAFT.PRT	1,261
CLAMP.PRT	7,928	WASHER.PTN	876
GASKET.PTN	4,791		

Part files _____

Pattern files _____

Unit 2 *SUBTRACTION OF WHOLE NUMBERS*

BASIC PRINCIPLES OF SUBTRACTION OF WHOLE NUMBERS

Subtraction is the process of finding the *difference,* or *remainder,* between two numbers. The smaller of the two numbers is placed below the larger, keeping the right column of numbers aligned.

Begin your work from the units column at the right.

Example: Subtract 523 from 867.

```
    867    (minuend)
  - 523    (subtrahend)
    344    (remainder or difference)
```

If the digit being subtracted (subtrahend) is larger than the top digit (minuend), a 1 is borrowed from the digit in the column to the left as shown below.

Example:
```
              ┌──── (8 becomes 7)
              ↓
        386  ← (6 becomes 16)
       -  57
        329
```

  867 ⊖ 523 ⊜ 344

PRACTICAL PROBLEMS

Subtract the following quantities.

1.   98 feet
    − 16 feet

2.   162 inches
    − 37 inches

3.   347 millimeters
    − 206 millimeters

4. 1,476 pounds
    − 808 pounds

5.   431 yards
    − 116 yards

6.   263 inches
    − 79 inches

7.   337 feet
    − 183 feet

8. 4,153 millimeters
    − 1,276 millimeters

9.   There are 365 days in a year. A certain drafter works 226 days. How many days is the drafter not working?   _____

10. What is the threaded length, in millimeters, of this round head machine screw?   _____

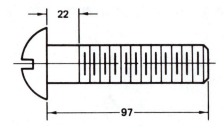

11. The GROUP command links entities on a drawing together so they can be selected as a complete unit. A CAD drawing contains 1,465 entities in one group. Seventeen entities are deleted from the front view, 21 from the top view, 37 from the side view, and 13 from the pictorial view. All are contained in the group. How many entities are left in the group after these specific entities are deleted?   _____

12. During one month, 236 working drawings are submitted to a company checking team. Only 187 drawings are checked and returned to the drafters. How many drawings are not yet returned?   _____

13. An assembly drawing has 76 parts. The first drafter details 14 parts. The second details 21 parts. The third details 17 parts. How many parts are not yet drawn?   _____

14. A CAD drafter's time limit for a specific job is 500 hours. The time already spent is 40 hr., 36 hr., 50 hr., 48 hr., 40 hr., 42 hr., 38 hr., and 44 hr. How many more hours can be spent on this job?   _____

15. A large company has 300 drafters. During the year 45 are laid off, 23 retire, 37 leave the company, and 8 are promoted out of the department. How many additional drafters need to be hired to maintain 250 drafters in the company?   _____

16. Find the overall length, in inches, of this T square. _____

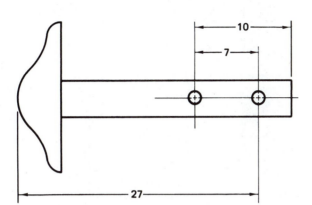

17. Find, in millimeters, distances **A, B,** and **C** on this template.

A _____

B _____

C _____

18. In a certification program there are 109 drafting students. One class has 28 pupils, the second class has 26 pupils, and another class has 27 pupils. How many pupils are in the fourth class? _____

19.   Use the top and front views of this figure to find, in millimeters, distances **A, B,** and **C.**

A _____

B _____

C _____

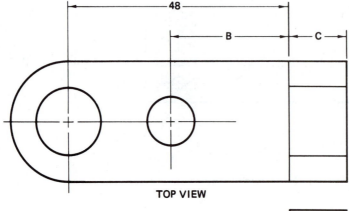

TOP VIEW

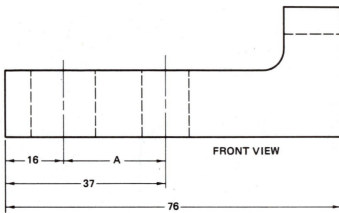

FRONT VIEW

20.   A drafter needs to find the missing length on a drawing. The overall length is 284 mm. The other lengths are measured as 17 mm, 35 mm, 67 mm, 27 mm, 52 mm, 26 mm, 11 mm, and 23 mm. Determine the missing length.

_____

21.   A CAD drawing of a large building contains 950 symbols made up of 110 windows, 72 doors, 33 appliances, 215 lighting fixtures, 365 electrical switches, and 89 mechanical equipment symbols. How many other symbols are shown on the drawing that have not been mentioned?

_____

22.   A large assembly drawing contains 96 different parts that need to be detailed. A team of CAD drafters has detailed the following number of parts: 7, 5, 11, 15, 9, 13, and 8. How many parts still need to be detailed?

_____

23. Using the plot plan showing property lines and setbacks, find the difference between the property perimeter and the setback perimeter if the property lines measure 185' × 105'.

_____

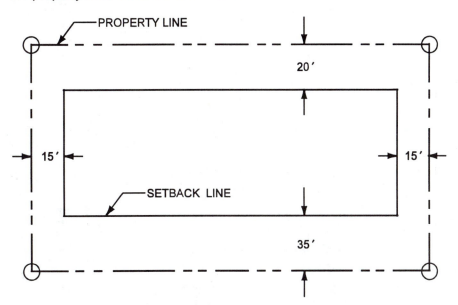

24. A CAD operator has 256 levels available to be used for drawing and plotting. Currently 126 are being used (active). Determine how many levels are still active if the following levels are turned off for plotting purposes.

_____

Construction Details	17	East Elevations	3
Foundation Details	5	West Elevations	3
Door Schedules	7	South Elevations	3
Window Schedules	6	Floor Plans	45

25. Find the difference between the property line perimeter and the perimeter of the house using the CAD drawing illustrated.   _____

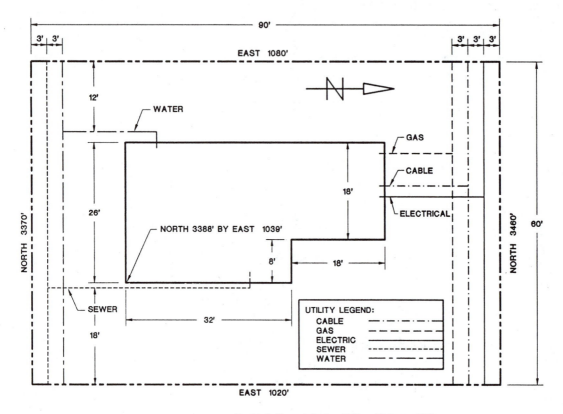

Reprinted with permission from Stellman, Krishnan, and Rhea,
Harnessing AutoCAD, copyright 1993 by Delmar Publishers.

26. A CAD drafter needs to know the overall width and height for placement purposes. Using the CAD drawing provided, find the overall height and width.

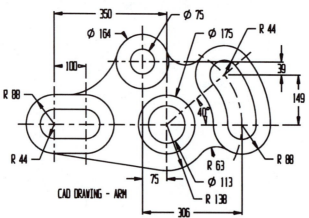

Height _____

Width _____

27. Using the CAD drawing below, find the size of radius **A**.   _____

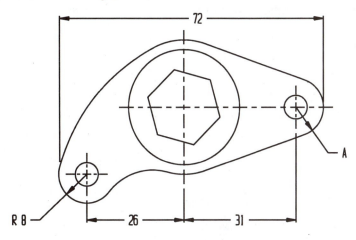

CAD DRAWING - ROCKER ARM

28. Determine the values for dimensions **A**, **B**, and **C** for the CAD drawing presented.

A _____

B _____

C _____

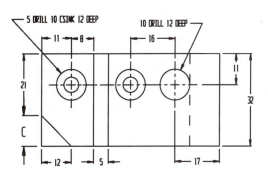

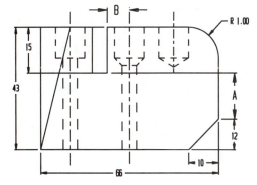

CAD DRAWING - LOCATING BLOCK

29. Find the difference between the sum of the perimeters of the house, garage, and driveway and the perimeter of the property lines. Use the CAD drawing presented.

_____

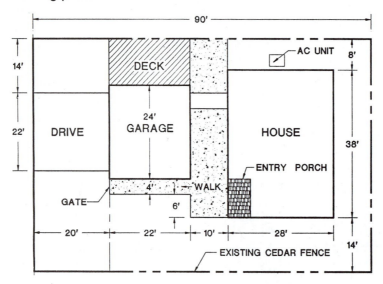

Reprinted with permission from Stellman, Krishnan, and Rhea, *Harnessing AutoCAD*, copyright 1993 by Delmar Publishers.

30. A floppy disk contains ten symbol files leaving 7,180 bytes still available for storage. A new symbol file having 22,012 bytes must be saved on this disk. What specific file(s) need to be deleted to allow the new symbol file to be saved, keeping the maximum number of bytes on the disk? Provide answers to the four blanks listed below.

FILES ON DISK

Symbol 1	47,050 bytes	Symbol 6	14,360 bytes
Symbol 2	35,751 bytes	Symbol 7	21,116 bytes
Symbol 3	74,362 bytes	Symbol 8	7,301 bytes
Symbol 4	62,171 bytes	Symbol 9	6,507 bytes
Symbol 5	89,753 bytes	Symbol 10	6,449 bytes

a. Current total number of bytes

_____

b. Maximum disk storage space (bytes)

_____

c. Specific file(s) to be deleted

_____

d. New maximum number of bytes

_____

Unit 3 MULTIPLICATION OF WHOLE NUMBERS

BASIC PRINCIPLES OF MULTIPLICATION OF WHOLE NUMBERS

Multiplication is actually a simple method of addition. For example, if four 6's are added, the answer will be 24. If the number 6 is multiplied by 4, the answer (known as the *product*) is equal to 24. Therefore, 6 × 4 is the same as adding four 6's.

```
        6
        6
        6                    6
      + 6                  × 4
       24                   24
```

To multiply more complex numbers, write the number to be multiplied. Then write under it the number of times it is to be multiplied. In the following example, the number 375 is to be multiplied by 24. Write the numbers keeping the right column aligned.

Example: 375 × 24

```
                                      1        11        11        11
     2          32          32        32        32        32        32
   375         375         375       375       375       375       375
 ×  24       ×  24       ×  24      ×  24     ×  24     ×  24     ×  24
     0          00        1500      1500      1500      1500      1500
                                       0        50       750     + 750
                                                                  9000
```

Multiply the first two numbers (4 × 5 =20). Place the 0 below the 4, and carry the two to the next column. Then multiply (4 × 7 = 28) and add the 2 (28 + 2 = 30). Place the zero beside the other zero and carry the three to the next column. Next, multiply (4 × 3 = 12) and add the 3 (12 + 3 = 15). Place the 15 beside the zero. Now multiply each number by the 2 in 24. The answers will be brought down in the same manner except that one space is skipped. Multiply (2 × 5 = 10). Place the zero below the correct zero and carry the 1 to the next column. Multiply (2 × 7 = 14). Add the 1 from the first column (14 + 1 = 15). Place the 5 beside the zero and carry the 1 to the next column. Then multiply (2 × 3 = 6) and add the 1 (6 + 1 = 7). Place the 7 beside the 5. The final step is to add the two sets of products together to obtain their sum.

📱 375 ⓧ 24 ⚌ 9000

PRACTICAL PROBLEMS

Multiply the following quantities.

1.   74   cm   2.   107   lb.   3.   345in.4.133hr.
         ×   7   ×   8   ×   12×   18

5.   17   ft.   6.   65   in.   7.   36yd.8.29mm
         × 13   × 37   ×   24×   48

9.   A detail drafter measures off eight line segments each 3 inches long.
     How long, in inches, is the line from beginning to end?   _____

10. Find, in millimeters, the total length of artgum needed to make 7 erasers.   _____

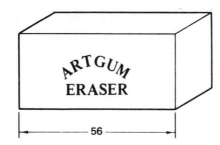

11. A CAD drafter works on an assembly drawing for eight hours in one day.
     The drafter estimates that it will take 6 more working days to complete
     the drawing. Find the total hours worked on this job if the estimate is
     correct. (A working day equals 8 hours.)   _____

12. Through a new manufacturing process, bar stock originally measuring
     157 mm can be shortened by 35 millimeters. If the new process is used,
     find the total length, in millimeters, of stock needed for 25 pieces.   _____

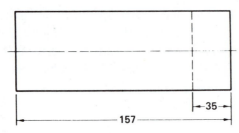

13. The end portion of this piece of brass stock is a regular hexagon. Find, in millimeters, the perimeter of the end portion. _____

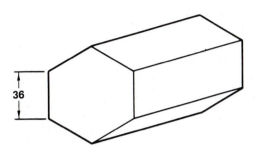

14. The *area* of a rectangle is found by multiplying the length times the width. What is the area, in square feet, of a drafting room 25' in width by 60' in length? _____

15. A drafting department supervisor orders 17 reams of 18" by 24" size paper. Each ream weighs 6 pounds. What is the total weight, in pounds, of this paper? _____

16. All holes on this drilled plate are equally spaced. Determine the total length, in millimeters, from the center of hole **A** to the center of hole **H**. _____

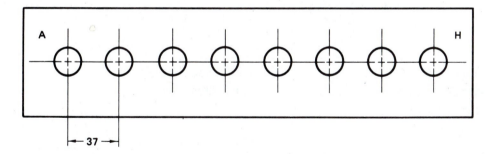

17. If there are 100 centimeters in 1 meter, how many centimeters long is this strap? _____

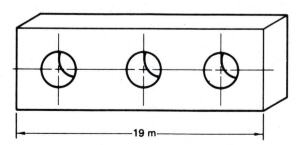

18. If there are 10 decimeters in 1 meter, how many decimeters are contained in the overall length of the connecting link?   _____

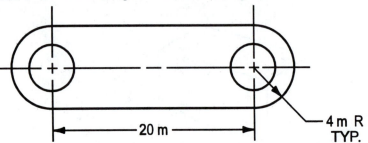

19. A CAD drafter must locate seven 1" holes at 35° intervals about a circle. Determine the number of degrees from the center of hole **1** to the center of hole **7**   _____

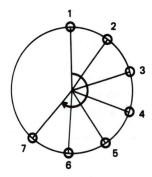

20. A *checker* is a person who carefully looks at drawings to see if they are correct. A certain checker examines an average of 7 drawings a day. How many drawings are checked in:

   a. 5 days?   _____

   b. 20 days?   _____

21. All sections of this sheet metal development are equal length except the tab. Find, in millimeters, the overall length including the tab. _____

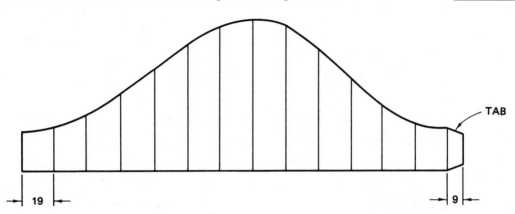

22. A CAD operator must figure out the total cost of the wall switches shown on a first floor plan. There are 17 single pole, single throw switches at $2.00; 19 single pole, double throw switches at $3.00; and 7 double pole, double throw switches at $4.00.

    a. Find the total cost of switches for the first floor. _____

    b. Using the total cost as an average cost, what would be the cost of switches for the remaining 14 floors? _____

    c. Find the cost of all switches mentioned for the entire building. _____

23. There are eight bits of information in every byte. How many bits are contained in a drawing file having 3,506 bytes? _____

24. CAD drafters use the scale command to increase or decrease the size of objects. Determine the new height and width of this CAD drawing if it is increased by a factor of 4.

Height _____

Width _____

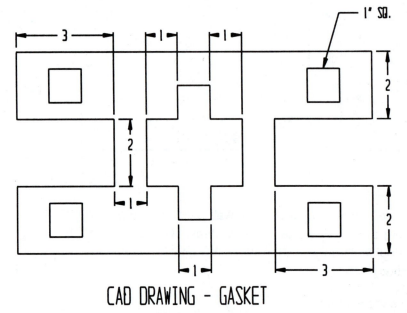

CAD DRAWING - GASKET

25. Using the CAD drawing below, determine the overall length and dimensions **A–E** if the arm is increased by a factor of 8 using the SCALE command.

A _____

B _____

C _____

D _____

E _____

Length _____

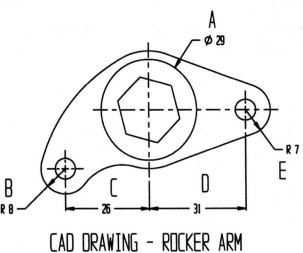

CAD DRAWING - ROCKER ARM

26. The CAD drawing illustrated will be scaled up by a factor of 3. Determine the new drive, garage, and house perimeters.

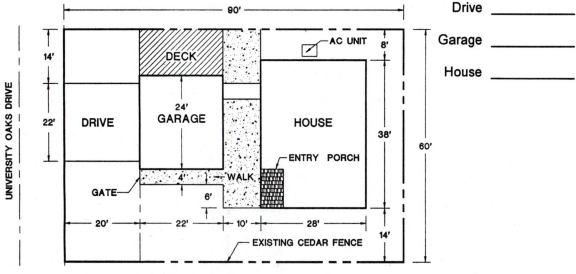

Drive _____

Garage _____

House _____

Reprinted with permission from Stellman, Krishnan, and Rhea,
Harnessing AutoCAD, copyright 1993 by Delmar Publishers.

27. The symbol shown using the absolute coordinate system will be scaled up by a factor of 7. Find its new height and width as well as the new diameter size.

Height _____

Width _____

Diameter _____

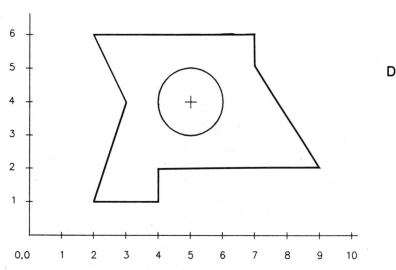

Reprinted with permission from McGrew,
Exploring the Power of AutoCAD, copyright 1990 by Delmar Publishers.

28. A CAD drafter must determine how many symbols are contained on a recently completed project. Levels 1, 3, and 7 each have seven symbols; 2, 5, and 9 each possess eleven symbols; 4, 8, and 16 each contain twelve; 6, 10, and 15 each have eight; and 11, 12, 13, and 14 each have nine symbols. Find the total number of symbols used with this project.

_____

29. The CAD drawing below will be scaled up by a factor of 9. Find the new sizes for dimensions **A–E** once the scaling operation is completed.

A _____

B _____

C _____

D _____

E _____

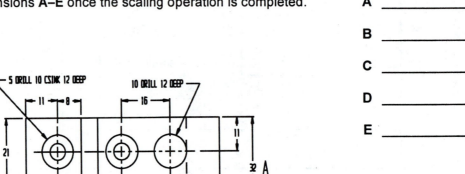

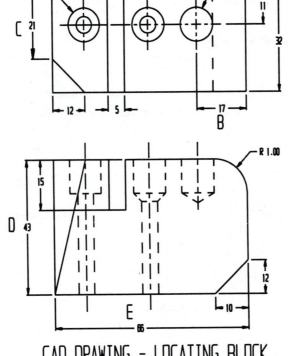

CAD DRAWING - LOCATING BLOCK

30. The CAD drawing of the arm is to be scaled up by a factor of (6X). Calculate the new sizes for dimensions **A–E**.

A _____

B _____

C _____

D _____

E _____

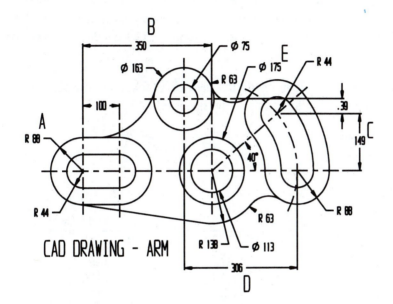

CAD DRAWING - ARM

Unit 4 DIVISION OF WHOLE NUMBERS

BASIC PRINCIPLES OF DIVISION OF WHOLE NUMBERS

In Unit 3, it was shown that multiplication is actually the process of adding a number together many times; division is just the opposite. Division is actually the process of subtracting a smaller number from a larger number many times. The number to be divided is referred to as the *dividend*. The number used to indicate the number of times the dividend is to be divided is called the *divisor,* and the answer is known as the *quotient*.

To begin the process, the dividend is placed inside the division bracket, the divisor is placed to the left of the dividend, and the quotient is placed above the dividend.

$$\text{Divisor} \overline{)\,\text{Dividend}}^{\displaystyle \text{Quotient}}$$

Example: Divide 1310 by 15.

```
       8              8              87             87  R5
15 ) 1310       15 ) 1310       15 ) 1310       15 ) 1310
     120            120            120            120
      11            110            110            110
                                   105            105
                                                    5
```

Place the number 1310 under the division bracket and the number 15 to the left of it. The number 15 cannot be divided into a number which is smaller than itself. Therefore, 15 is divided into the number 131 first. The first step is to find what number when multiplied by 15 will come the closest to 131 without going over 131. In this example, 8 is that number (8 × 15 = 120). The number 120 is placed below 131 and the difference is found. This leaves 11. 15 cannot be divided into 11, so the next number is brought down to the right of 11. When the zero is placed beside 11, 15 is then divided into 110. 7 is found to be the correct multiplier (7 × 15 = 105). 105 is placed below 110 and a difference of 5 is found. Because there are no more numbers in the dividend, the 5 is placed in the quotient and shown as R5 which means a "Remainder" of 5.

▤   1310 ÷ 5 = 87.33

PRACTICAL PROBLEMS

Divide the following quantities.

1. $6\overline{)384}$ in.   _____

2. $783 \div 9$ _____

3. $13\overline{)819}$ cm   _____

4. $1404 \div 27$ _____

5. $116\overline{)4{,}176}$ mm   _____

6. $391 \div 17$ _____

7. An architectural drafter lays out (measures off) a line 72 inches long. The line is divided into 12-inch lengths. How many sections are obtained from this line?   _____

8. A note on a working drawing states that 6 holes are equally located around a circle. How many degrees are in each equal space between the holes?   _____

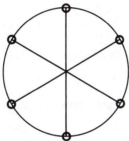

9. How many 9-inch line segments are there in a straight line 108 inches long?   _____

10. How many 15-degree sections are there in the circle illustrated?   _____

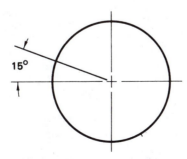

11. Six CAD drafters spend a total of 498 hours working on a special job. Each one works the same number of hours. How many hours does each one work?

_____

12. On this block, the distance between the centers of the holes is the same as the distance from the end to the center of the hole nearest the end. Determine the distance, in millimeters, between centers.

_____

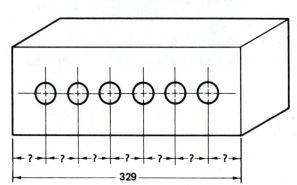

13. Forty drafting boards weigh 280 pounds. How many pounds does each board weigh?

_____

14. How many degrees are there between each hole location in this illustration?

_____

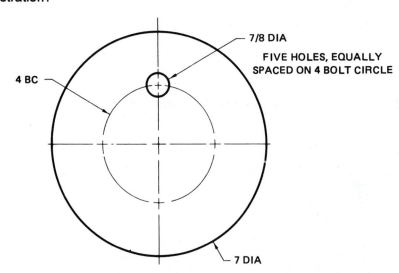

15. Three dozen erasing shields cost 12 dollars. What is the cost of 1 dozen erasing shields?

_____

16. How many 4-inch erasing shields can be cut from a strip of sheet metal 128 inches long? (No allowance is made for waste.) _____

17. The radius of a circle is found by dividing the diameter by 2. What is the radius of a 134-mm diameter circle? _____

18. In 7 hours, 5,040 triangles are made. How many are made in:

    a. one hour? _____

    b. one minute? _____

19. A package of 2H drafting pencils contains 12 pencils. How many packages are made from a box containing 720 pencils? _____

20. All sections of this template are of equal width. Determine the width, in millimeters, of each section. _____

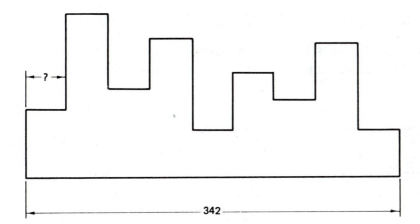

21. Determine how many bytes are contained in a drawing file containing 38,312 bits of information. _____

22. Five drawing files contain the following number of bytes: 12,337; 21,142; 10,615; 26,457; and 19,579. Find the average file size in bytes. _____

23. The CAD drawing shown is to be reduced by a factor of 3. Find the resulting height and width once the scaling operation is performed. Note: Grid spacing equals 8 inches.

Height _____

Width _____

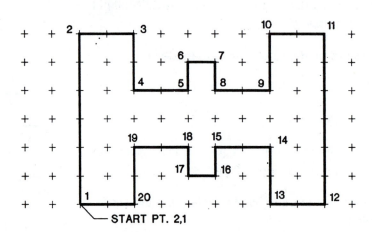

START PT. 2,1

Reprinted with permission from Stellman, Krishnan, and Rhea,
Harnessing AutoCAD, copyright 1993 by Delmar Publishers.

24. The CAD drawing of the house is to be scaled down by a factor of 4. How long and how high will the house be once the scaling operation is completed?

Length _____

Height _____

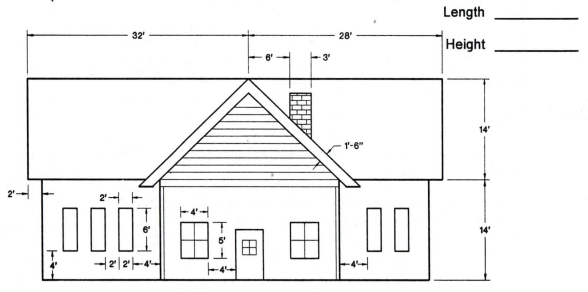

Reprinted with permission from Stellman, Krishnan, and Rhea,
Harnessing AutoCAD, copyright 1993 by Delmar Publishers.

25. A CAD drafter uses the SEGMENT command to divide a line into equal segments. Determine the segment length if line **A** is 119 mm long with 17 segments and line **B** is 78 mm long and is divided into 13 segments.   A _____

   B _____

26. A CAD operator wishes to set up the 256 levels used for drawing so that eight different types of drawings may have the same number of levels. Using this procedure, how many levels are assigned to each drawing type?   _____

27. The DIVIDE command divides a line into a specific number of divisions and places a symbol at each division. A property line on a plot plan that is 180 feet long shows twelve symbols. Determine the spacing in feet between symbols.   _____

28. The CAD drawing shown must be scaled down by a factor of 8. Find the overall length after the arm has been reduced.   _____

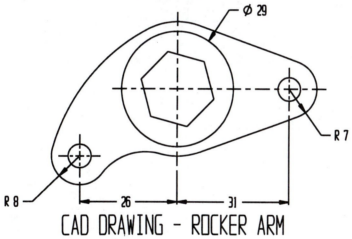

CAD DRAWING - ROCKER ARM

29. Currently 120 levels are active on a CAD project. If this number is reduced by half and then by half once more, how many levels are now active?   _____

30.   The CAD drawing of the arm will be reduced by a factor of 12. Determine the new overall height and width.

Height _____

Width _____

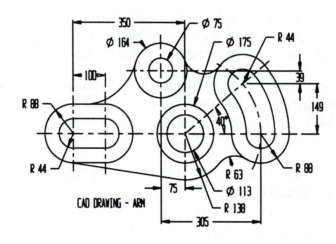

Unit 5 COMBINED OPERATIONS WITH WHOLE NUMBERS

BASIC PRINCIPLES OF COMBINED OPERATIONS

This unit contains practical problems involving combined operations of addition, subtraction, multiplication, and division with whole numbers.

PRACTICAL PROBLEMS

1. A *spacing collar* is used to separate adjacent parts when assembled. Find, in inches, the wall thickness of this spacing collar. Note: The wall thickness is the same on each side.

   _____

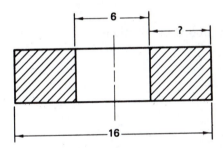

2. A drafting student centers a 4" by 6" block on an 8" by 12" sheet of paper. Find, in inches, dimensions **A** and **B**.

   A _____

   B _____

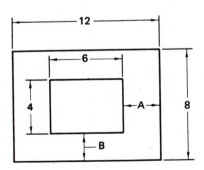

3. What is the perimeter of this shim in millimeters?   _____

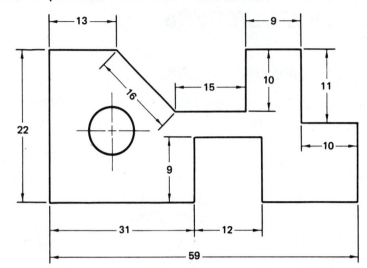

4. Find, in millimeters, dimensions **A, B,** and **C** of this plate.

A _____

B _____

C _____

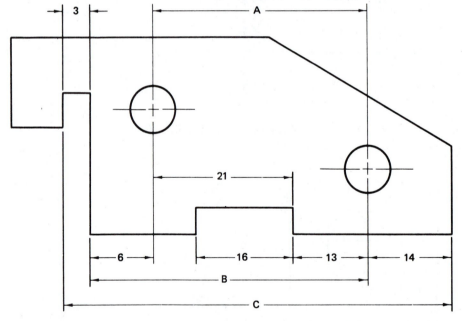

5. The number of T squares assembled on each shift is 436, 427, and 412.

    a. How many T squares were made?   _____

    b. Each shift produces 16 defective T squares. How many products
        were made to specifications?   _____

6. A circle contains 12 equal sections. How many degrees are in each section? _____

7. Square hole **A** is punched into surface **B**. Surface **B** is 441 square inches before hole **A** is punched. Find, in square inches, the surface area remaining after hole **A** is punched. _____

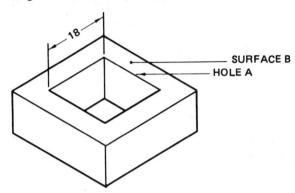

SURFACE B
HOLE A

8. Sixteen CAD drafters work for 5 days. Twenty-five architectural drafters work for 8 days. Each workday is 8 hours long. What is the total number of hours they work? _____

9. What is the total area, in square inches, of surface **A** and surface **B**? _____

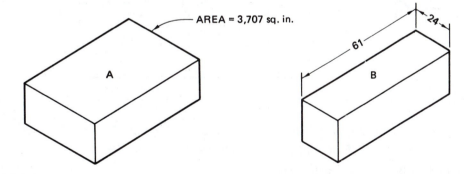

AREA = 3,707 sq. in.

A

61

24

B

10. Two departments complete a package of working drawings in 1,856 man-hours. The 9 drafters in Department *X* and the 7 drafters in Department *Y* each worked an equal number of hours. How many more hours did Department *X* work? _____

11. On this gauge, distance **AB** is equal to distance **CD**. Find, in millimeters, the distance **AB**.

_____

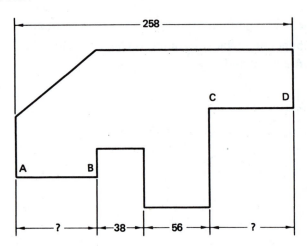

12. Determine length **X**, in millimeters, on the development shown. All sectional distances are equal.

_____

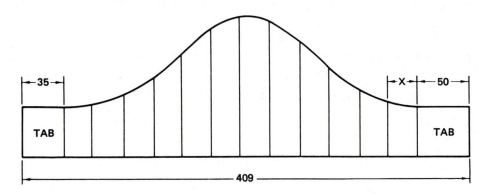

13. An engineering department employs 268 drafters in divisions X and Z. Each engineer is assigned four drafters. Department X has 43 engineers. How many engineers are in division Z?

_____

14. A *regular octagon* has eight equal sides. What is the difference, in millimeters, between these perimeters? _____

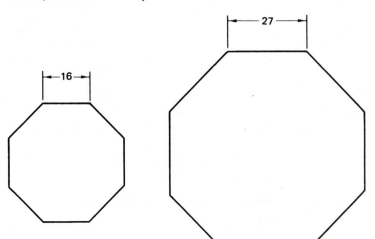

15. A piece of shafting is 72" long. Lengths of 8", 13", 21", 5", and 19" are cut from it. Determine how much material is left. No allowance is made for material lost in the cutting process. _____

16. Twenty machine screws are made in one minute. Determine how many are made in one hour and forty-five minutes and fifteen seconds. _____

17. What is the total length of 66-mm wide strip needed to make 120 T guides? State answer in millimeters. _____

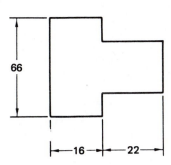

18. What is the total area, in square inches, of the two pieces of sheet
metal?   _____

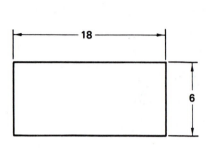 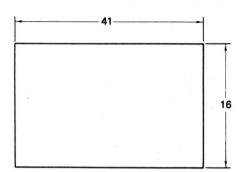

19. Determine the difference between the area of surface **A** and the area of
surface **B**. State answer in square inches.   _____

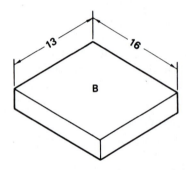 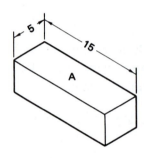

20. Determine distances **A, B, C,** and **D,** in inches, from the top and front
views of the wedge block.

A _____

B _____

C _____

D _____

21. CAD drafters make electronic drawings on transparent overlays called levels or layers. Many CAD systems contain 256 separate levels that may be used independently. A specific CAD drawing uses 28 levels for construction details, 61 for floor plans, 18 for elevations, 16 for electrical symbols, 22 for furniture and appliances, and 49 for dimensions.

   a. What is the total number of levels already assigned a specific function? _____

   b. How many levels from the 256 available have not been assigned a specific function? _____

22. The front view of the object shown using absolute coordinates must be increased by a factor of 5. Find the new height and width. The coordinate spacing is in one-inch intervals.

   Height _____

   Width _____

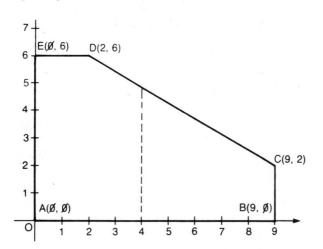

Reprinted with permission from Zandi,
Computer-Aided Drafting and Design, copyright 1985 by Delmar Publishers

23. A floppy disk has 272,000 bytes left for file storage. Find the space (number of bytes) remaining after the following part files are saved to the disk: 36,945; 71,541; 17,612; 56,744; 27,945; and 7,345. _____

24. How many bits of information are contained in the following four files: 72,396; 15,819; 56,371; and 32,154? _____

25. The CAD drawing of the house is to be scaled down by a factor of 5 on the *X*-axis (length) and by a factor of 4 on the *Y*-axis (height). Determine the new height and length resulting from the scaling process.

Height _____

Length _____

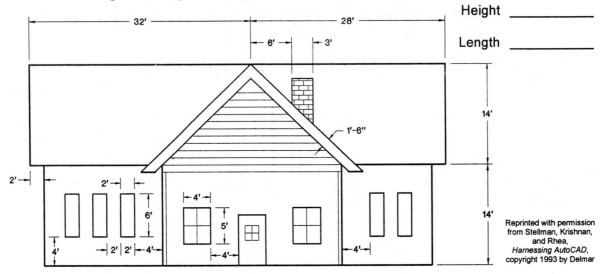

Reprinted with permission from Stellman, Krishnan, and Rhea, *Harnessing AutoCAD*, copyright 1993 by Delmar

26. Determine the new overall height, width, and depth dimensions if the CAD drawing is enlarged by a factor of 6.

Height _____

Width _____

Depth _____

Reprinted with permission from Zandi, *Computer-Aided Drafting and Design*, copyright 1985 by Delmar Publishers.

27. Using the CAD drawing presented, find the new sizes for features **A, B,** and **C** after they are scaled up by a factor of 8.

A _____

B _____

C _____

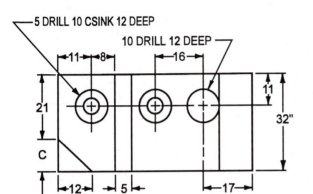

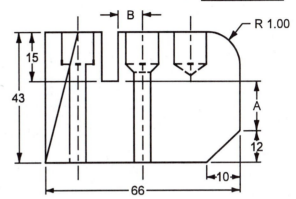

28. The STRETCH command used in CAD allows the CAD operator to stretch the shape of an object without affecting other crucial parts. In the CAD drawing presented, the object's length will be enlarged (stretched) by twice the distance between holes **A** and **B**. Determine the new overall length resulting from this operation.

_____

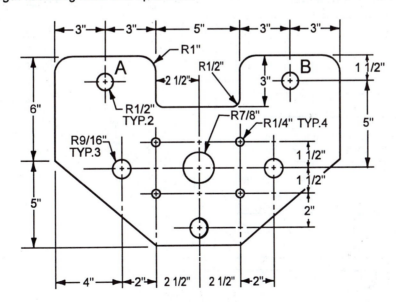

29. The CAD drawing illustrated will be scaled down by a factor of 7.
     Determine the new line lengths for **AB**, **BC**, **FG**, **GH**, **HI**, and **AI**.

AB _____

BC _____

FG _____

GH _____

HI _____

AI _____

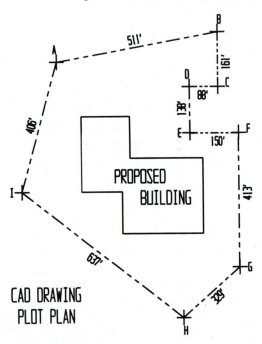

CAD DRAWING
PLOT PLAN

30. Using the CAD drawing below, find dimensions **A–E**.

A _____

B _____

C _____

D _____

E _____

CAD DRAWING – RETAINER

31. Using the CAD drawing shown in isometric, find the new sizes for **A, B,** and **C** if the view will be scaled down by a factor of 7.

A _____

B _____

C _____

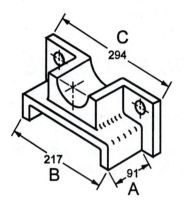

Common Fractions

Unit 6   ADDITION OF COMMON FRACTIONS

BASIC PRINCIPLES OF ADDITION OF COMMON FRACTIONS

Accurate measurement in all drafting fields requires closer measurement than can be accomplished using only whole numbers. This can be done by dividing the inch into many equal parts called common fractions:

$$\tfrac{1}{2}, \tfrac{1}{4}, \tfrac{1}{8}, \tfrac{1}{16}, \tfrac{1}{32}, \text{ and } \tfrac{1}{64}.$$

There are two parts to a common fraction, the numerator (the number above the line) and the denominator (the number below the line). The denominator indicates the number of equal parts the whole is divided into. To add or subtract fractions, you must first find the least common denominator (LCD). The least common denominator is the smallest number that all of the denominators will divide into.

Example: $\tfrac{1}{2}, \tfrac{1}{4}, \tfrac{1}{8}, \tfrac{1}{16}, \tfrac{1}{64}$

   LCD = 64

When adding fractions, all denominators must be the same. If so, only the numerators are added.

Example: $\tfrac{3}{16} + \tfrac{5}{16} + \tfrac{7}{16}$

$$
\begin{array}{r}
\tfrac{3}{16} \\
\tfrac{5}{16} \\
+\ \tfrac{7}{16} \\
\hline
\tfrac{15}{16}
\end{array}
$$

When the denominators are not the same, it is necessary to find the common denominator. The common denominator may always be found by multiplying all denominators.

Example: $\tfrac{1}{2} + \tfrac{1}{3} + \tfrac{1}{8}$ (2 × 3 × 8 = 48)

This often results in a large number and a lower one may be found by inspection. In this case, 24 would be the lowest common denominator.

Adding these three fractions with different denominators is accomplished in the following manner:

1. Divide the LCD (24) by the denominator of each fraction.

2. Multiply both numerator and denominator of each fraction.

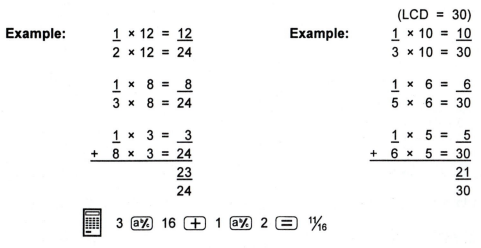

Example:

$$\frac{1}{2} \times \frac{12}{12} = \frac{12}{24}$$

$$\frac{1}{3} \times \frac{8}{8} = \frac{8}{24}$$

$$+ \ \frac{1}{8} \times \frac{3}{3} = \frac{3}{24}$$

$$\frac{23}{24}$$

Example: (LCD = 30)

$$\frac{1}{3} \times \frac{10}{10} = \frac{10}{30}$$

$$\frac{1}{5} \times \frac{6}{6} = \frac{6}{30}$$

$$+ \ \frac{1}{6} \times \frac{5}{5} = \frac{5}{30}$$

$$\frac{21}{30}$$

3 [a%] 16 [+] 1 [a%] 2 [=] $\frac{11}{16}$

When working in related drafting fields you will encounter measurements given in feet and inches.

When converting feet and inches to inches, multiply the number of feet by 12, then add the number of inches to the product.

Example: Express 3 feet 7 $\frac{3}{4}$ inches as inches. (3'–7 $\frac{3}{4}$ ")
3 × 12 = 36   3 feet = 36 inches
36 inches plus 7 $\frac{3}{4}$ inches = 43 $\frac{3}{4}$ inches

To express inches as feet and inches, divide the number of inches by 12 to find the number of feet. The remainder equals the number of inches.

Example: Express 145 $\frac{1}{2}$ inches as feet and inches.

$$\begin{array}{r} 12 \ \text{(feet)} \\ 12\,)\,\overline{145 \ \tfrac{1}{2}} \\ \underline{12} \\ 25 \\ \underline{24} \\ 1 \ \tfrac{1}{2} \ \text{Remainder (inches)} \end{array}$$

Answer: 12'–1 $\frac{1}{2}$ "

PRACTICAL PROBLEMS

Find the sum of the following. Express all answers in lowest terms.

1. $\frac{3}{8}$ inch + $\frac{5}{8}$ inch =   _____

2. $\frac{1}{4}$ inch + $\frac{1}{2}$ inch =   _____

3. $\frac{5}{16}$ inch + $\frac{7}{16}$ inch =   _____

4. $\frac{1}{8}$ inch + $\frac{3}{16}$ inch =   _____

5. $2\frac{1}{4}$ inches + $3\frac{3}{16}$ inches + $4\frac{5}{8}$ inches =   _____

6. $3\frac{1}{2}$ hours + $2\frac{1}{4}$ hours + $1\frac{3}{4}$ hours =   _____

7. $4\frac{1}{3}$ hours + $\frac{1}{2}$ hour + $3\frac{5}{6}$ hours =   _____

8. $2\frac{7}{16}$ inches + 1 inch + $3\frac{17}{32}$ inches =   _____

9. $6\frac{2}{3}$ yards + $4\frac{1}{2}$ yards + $1\frac{1}{4}$ yards =   _____

10. $5\frac{11}{16}$ inches + 3 inches + $1\frac{47}{64}$ inches =   _____

11. Find, in inches, the length of this flat head cap screw.   _____

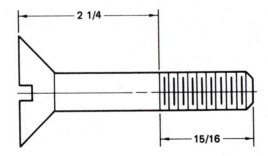

12. What is the length, in inches, of this link?   _____

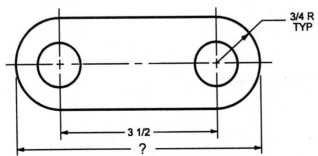

13. Determine the total length of a line formed by measurements of $2\frac{1}{8}$ ", $1\frac{3}{4}$ ", $3\frac{1}{2}$ ", and $1\frac{3}{8}$ ".

_____

14. Find, in inches, the overall length of a shaft made up of five sections: $3\frac{5}{16}$ ", $2\frac{7}{8}$ ", $1\frac{3}{4}$ ", $1\frac{9}{32}$ ", and $4\frac{1}{2}$ ".

_____

15. What are the lengths, in inches, of dimension **A** and dimension **B** on this template?

A _____

B _____

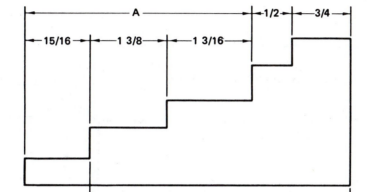

16. When $4\frac{3}{8}$ ", $3\frac{1}{16}$ ", $2\frac{9}{16}$ ", and $3\frac{1}{2}$ " are added, what is the total length?

_____

17. Calculate, in inches, the total length of this shaft.

_____

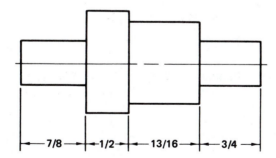

18. What length of stock, in inches, is needed to make a shaft with linear dimensions of $\frac{15}{32}$ ", $1\frac{3}{16}$ ", $2\frac{7}{32}$ ", $3\frac{9}{16}$ ", and $\frac{3}{4}$ "?

_____

19. Determine dimensions **A** and **B**, in inches, on this strap.

A _____

B _____

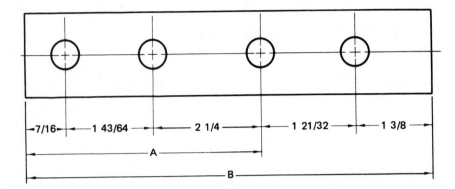

20. Find, in inches, dimensions **A** and **B** on this locator block.

A _____

B _____

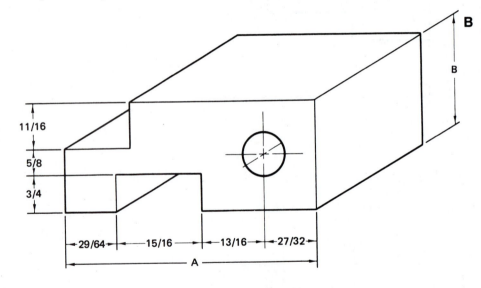

21. Using the diagram presented, find dimension **F**, if **A** is $^{11}/_{16}$, **G** is $3\frac{5}{8}$, and **E** is $2\frac{27}{32}$.

_____

22. Usually property line lengths are given in decimal feet. Determine the perimeter of the plot plan using the figures as presented.   _____

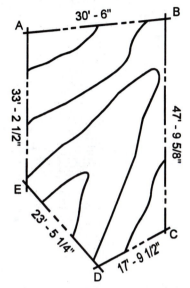

23. Using the CAD drawing below, find the overall length of the arm in inches.   _____

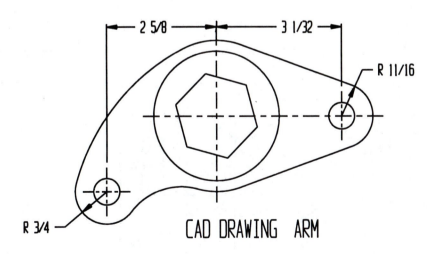

CAD DRAWING ARM

24.   Using the CAD drawing and dimensions given, find the perimeter of the deck, bedroom, and living room. Express your answer in feet and inches.

Deck _____

Bedroom _____

Living room _____

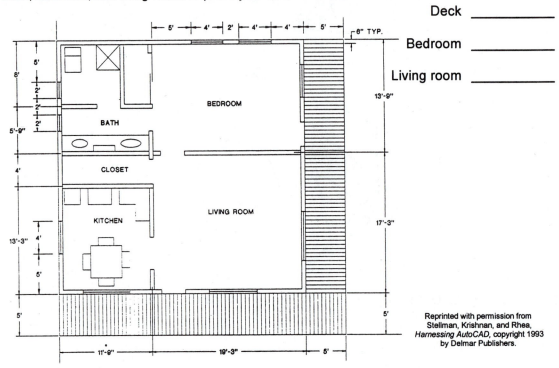

Reprinted with permission from Stellman, Krishnan, and Rhea, *Harnessing AutoCAD,* copyright 1993 by Delmar Publishers.

25.   Find dimensions **A** and **B** using the CAD drawing of the link. Express your answer in inches.

A _____

B _____

CAD DRAWING - LINK

26. Using the CAD drawing presented, find the height and width in inches.

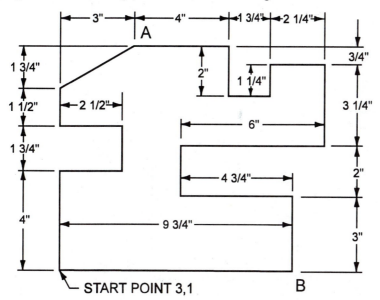

Height _____

Width _____

Reprinted with permission from Stellman, Krishnan, and Rhea,
Harnessing AutoCAD, copyright 1993 by Delmar Publishers.

27. Using the CAD drawing in problem 26, find the sum of line lengths from point **A** to point **B** in inches, moving in a clockwise direction. _____

28. The figure below is represented using the absolute coordinate system. Find the overall height and width if the unit spacing is $\frac{7}{16}$".

Height _____

Width _____

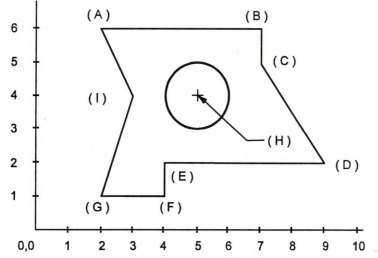

Reprinted with permission from McGrew,
Exploring the Power of AutoCAD, copyright 1990 by Delmar Publishers.

29.   Find dimension **A** in inches using the CAD drawing of the cam.   _____

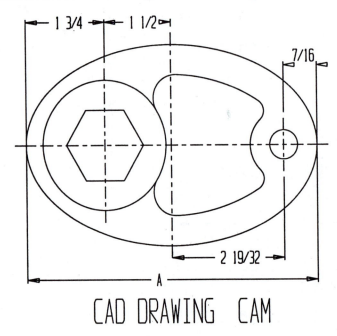

CAD DRAWING CAM

30.   Determine dimension **F** in inches if **A** is $\frac{7}{16}$", **B** is $\frac{47}{64}$", **C** is 1$\frac{3}{4}$", and   _____
      E is 2$\frac{3}{8}$".

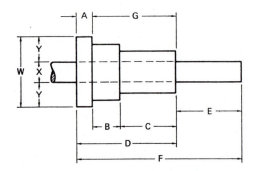

Unit 7 SUBTRACTION OF COMMON FRACTIONS

BASIC PRINCIPLES OF SUBTRACTION OF COMMON FRACTIONS

As in addition, fractions must be expressed with common denominators before subtracting. Once the common denominator is found, the numerator of the smaller fraction is subtracted from the larger numerator.

Example: $\dfrac{7}{8} - \dfrac{19}{32}$ (LCD = 32) $\dfrac{7}{8} = \dfrac{28}{32}$

$- \dfrac{19}{32} = \dfrac{19}{32}$

Answer: $\dfrac{9}{32}$

Example: $\dfrac{15}{16} - \dfrac{3}{4}$ (LCD = 16) $\dfrac{15}{16} = \dfrac{15}{16}$

$- \dfrac{3}{4} = \dfrac{12}{16}$

Answer: $\dfrac{3}{16}$

7 (a%) 8 (−) 19 (a%) 32 (=) $\frac{9}{32}$

PRACTICAL PROBLEMS

Subtract the following quantities. Express all answers in lowest terms.

1. $\dfrac{3}{4}$ inch
   $- \dfrac{3}{8}$ inch

2. $10\dfrac{3}{8}$ lb.
   $- 3$ lb.

3. $1\dfrac{7}{8}$ yd.
   $- \dfrac{5}{16}$ yd.

4. 21 ft.
   $- \dfrac{2}{3}$ ft.

5. $\dfrac{15}{16}$ inch $- \dfrac{15}{64}$ inch = _____

6. $4\dfrac{3}{8}$ lb. $- 1\dfrac{1}{2}$ lb. = _____

7. $3\dfrac{1}{4}$ hr. $- 1\dfrac{2}{3}$ hr. = _____

8. $2\dfrac{13}{64}$ inches from $3\dfrac{5}{8}$ inches = _____

9. $\dfrac{1}{8}$ foot from $\dfrac{3}{4}$ foot = _____

10. $1\dfrac{1}{2}$ pounds from $4\dfrac{3}{8}$ pounds = _____

11. A steel block is $\frac{27}{64}$ inch thick. A $\frac{3}{16}$-inch cut is taken on a milling machine. How thick, in inches, is the remaining block?

_____

12. A $\frac{3}{32}$-inch cut is taken from a $\frac{27}{64}$-inch block. What is the resulting thickness, in inches?

_____

13. A CAD drafter can spend 80 hours on three projects. If $61\frac{3}{4}$ hours are used, how many more hours are left to complete the projects?

_____

14. Find, in inches, the wall thickness of this collar.

_____

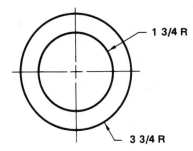

1 3/4 R

3 3/4 R

15. Each hole on this plate has a radius of $\frac{3}{4}$ inch. Find, in inches, dimension **A**.

_____

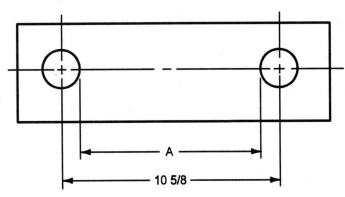

A

10 5/8

16. Find, in inches, the inside diameter of this washer.

_____

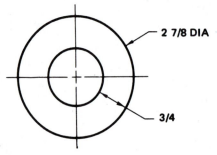

2 7/8 DIA

3/4

17. Find, in inches, dimensions **C** and **D** on this shim.

C _____

D _____

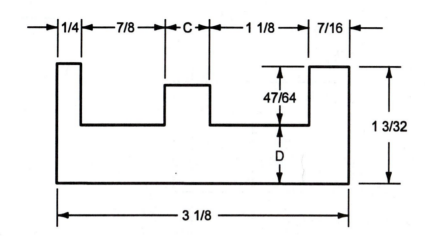

18. Find, in inches, dimension **A** on this template. Find the difference between the vertical dimensions.

A _____

Difference _____

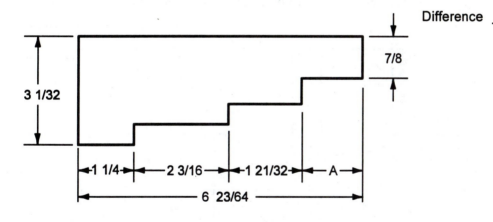

19. Three measurements on a drawing should total 7⅞". The actual measurements are 1½", 3⅛", and 2¹¹⁄₁₆". What is the difference between the actual and the required measurements?

_____

20. Calculate, in inches, dimensions **A** and **B** on this strap.

A _____

B _____

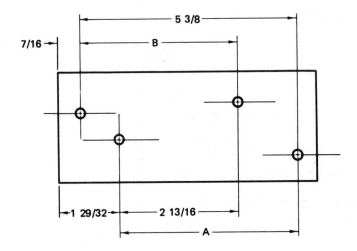

21. Find the depth of the drilled hole shown. Also, if a hole is first drilled to a depth of 1 $^{47}\!/_{64}$, how much deeper must it be drilled to reach the depth indicated?

a. _____

b. _____

22. What is the difference in diameters between 5 $\frac{1}{4}$ -inch and 3 $\frac{1}{2}$ -inch floppy diskettes?

_____

23. Using the diagram below, find, in inches, dimension **E** if **A** is ⅜ , **B** is 1 ¹⁄₁₆ , **C** is 3 ¹³⁄₃₂ , and **F** is 9 ⁴¹⁄₆₄ .

_____

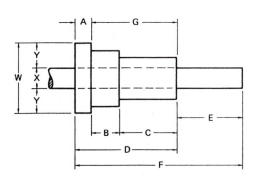

24. Using the CAD drawing below, find dimensions **A, B,** and **C**.

A _____

B _____

C _____

1ST FLOOR PLAN

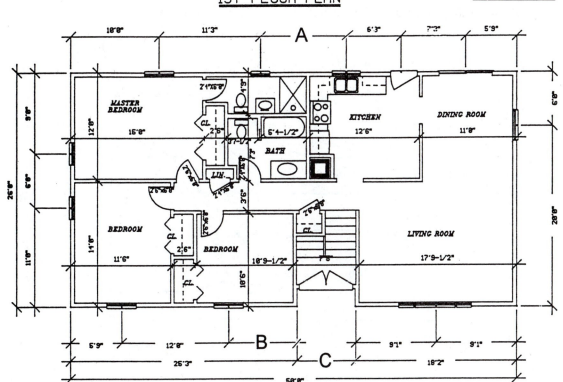

25. Find dimensions **A** and **B** on the CAD drawing provided.

A _____

B _____

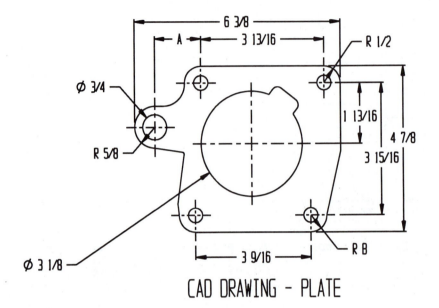

CAD DRAWING - PLATE

26. Calculate dimensions **A** and **B** using the CAD drawing illustrated.

A _____

B _____

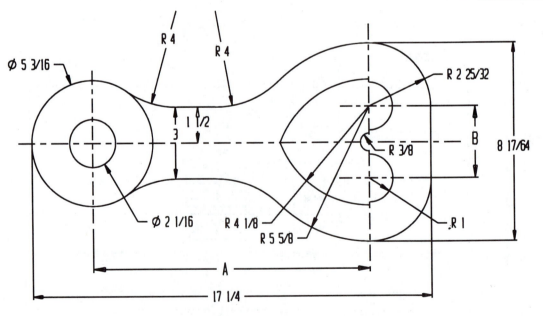

CAD DRAWING - LINK

27. Using the CAD drawing of the gasket, find the sizes of features **A** and **B**. **A** _____

    B _____

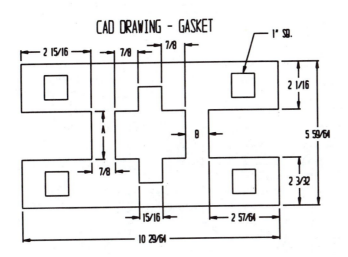

28. In the CAD drawing showing a wall section, find dimension **A** if the height from the bottom of the footing to the top of the subfloor is 2'-8 ¾". _____

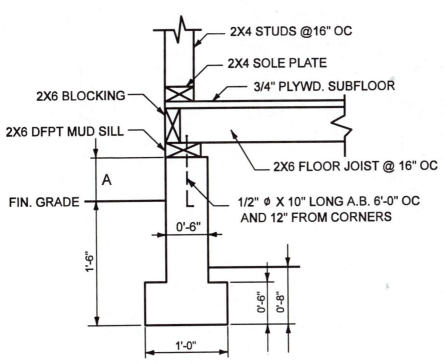

29. Find dimensions **A, B,** and **C** on the CAD drawing of the bracket.

A _____

B _____

C _____

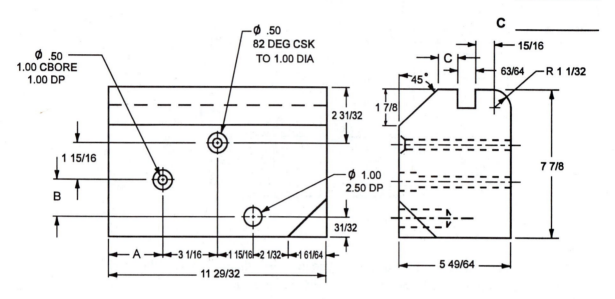

30. Using the CAD drawing provided, find dimensions **A** and **B**.

A _____

B _____

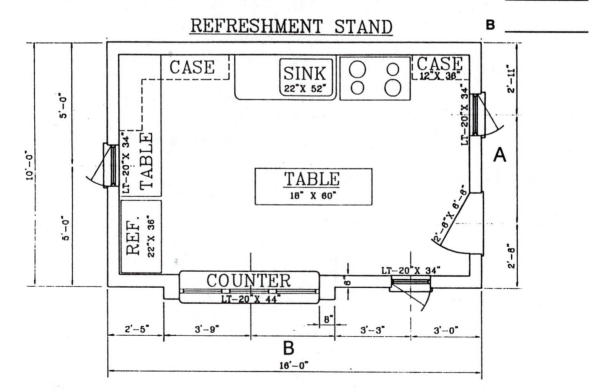

Unit 8 MULTIPLICATION OF COMMON FRACTIONS

BASIC PRINCIPLES OF MULTIPLICATION OF COMMON FRACTIONS

When multiplying two or more common fractions, multiply the numerators together and the denominators together, write the product of the numerators over the product of the denominators, reducing the fraction to the lowest terms.

Example: Multiply $\frac{3}{8}$ by $\frac{4}{5}$. $\frac{3 \times 4}{8 \times 5} = \frac{12}{40} = \frac{3}{10}$

The process can be simplified by *cancellation*. Before multiplying, divide the numerator and denominator by a number that is common to both.

Example:

$$\frac{3}{\overset{}{\underset{2}{\cancel{8}}}} \times \frac{\overset{1}{\cancel{4}}}{5} = \frac{3 \times 1}{2 \times 5} = \frac{3}{10}$$ (See above.)

When multiplying fractions by whole numbers, change the whole number to a fraction.

Example: $\frac{3}{16} \times 7 = \frac{3}{16} \times \frac{7}{1} = \frac{3 \times 7}{16 \times 1} = \frac{21}{16} = 1\frac{5}{16}$

When multiplying fractions by mixed numbers, change the mixed number to an improper fraction.

Example: $\frac{3}{4} \times 2\frac{1}{2} = \frac{3}{4} \times \frac{5}{2} = \frac{3 \times 5}{4 \times 2} = \frac{15}{8} = 1\frac{7}{8}$

🖩 3 (a%) 4 (X) 5 (a%) 2 (=) 1⅞

PRACTICAL PROBLEMS

Multiply the following quantities. Express all answers in lowest terms.

1. $\frac{3}{8} \times \frac{1}{2} =$ _____

2. $\frac{3}{16} \times \frac{1}{8} =$ _____

3. $\frac{1}{4} \times \frac{7}{16} =$ _____

4. $9 \times \frac{1}{2} =$ _____

5. $1\frac{5}{8} \times \frac{1}{3} =$ _____

6. $\frac{2}{3}$ by $\frac{3}{7} =$ _____

7. $2\frac{1}{2}$ by $8 =$ _____

8. $3\frac{1}{2}$ by $1\frac{1}{4}$ by $1\frac{3}{8} =$ _____

9. $1\frac{1}{2}$ by $\frac{3}{4}$ by $3\frac{1}{8} =$ _____

10. Seven pieces of copper are to be sheared from one length. Each piece is $2\frac{3}{4}$ inches long. What is the length of copper needed if there is no waste? _____

11. Cap screws are made from $\frac{3}{8}$" diameter bar stock. Each cap screw is $1\frac{3}{4}$" long. What length bar is needed to make 24 cap screws? _____

12. Find, in inches, the overall length of this sheet metal development. _____

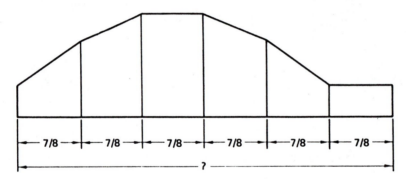

13. The thickness of a washer is $\frac{3}{32}$ inch. Calculate the height of a stack of 24 washers. _____

14. All holes on this plate are equally spaced. Determine the distance, in inches, between the centers of hole **A** and hole **B**. _____

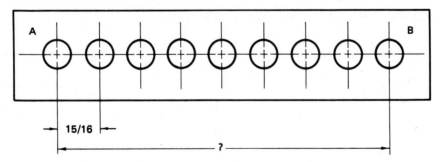

15. Drafting vellum costs 7 ½ cents a sheet. How much do 86 sheets cost? _____

16. The area of a rectangle is found by multiplying its length by its width. What is the area of this rectangle in square inches? _____

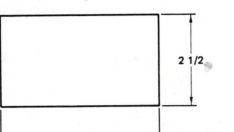

17. Calculate, in inches, the distance between floors and the length of dimension **A**.

Floors _____

A _____

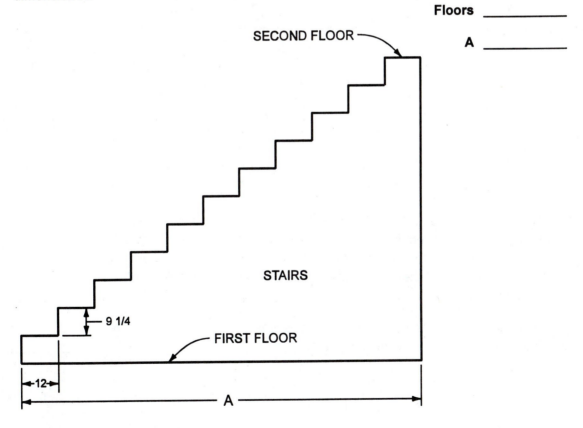

18. A box of 12-inch drafting scales weighs 3 ¾ pounds. A dealer has 20 ½ boxes in stock. What is the total weight of the scales? _____

19. An isometric projection is about ¾ the size of an isometric drawing. A 6-inch line is drawn on an isometric drawing. How many inches long is the line on the isometric projection?

_____

20. A CAD drafter lays off 12 line segments of 1¾ inches each. Calculate the total length, in inches, of these line segments.

_____

21. A regular hexagon has six equal sides and angles. Determine the perimeter, in inches, of this regular hexagon.

_____

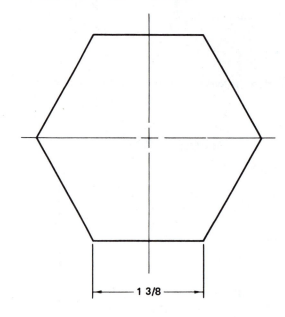

1 3/8

22. A CAD drafter spends a total of 9¼ work days on an architectural design project. During five work days, 9½ hours are devoted each day to the project, and for the remaining work days, 8½ hours are devoted to the project. Find the total number of hours that were devoted to this assignment.

_____

23. If ¼ inch on an architectural drawing represents 1 foot, how many inches on the drawing represent 39 feet?

_____

24. The view shown using absolute coordinates will be scaled up by a factor of 4. Determine the new height and width dimensions. The current coordinate spacing is ⅜".

Height _____

Width _____

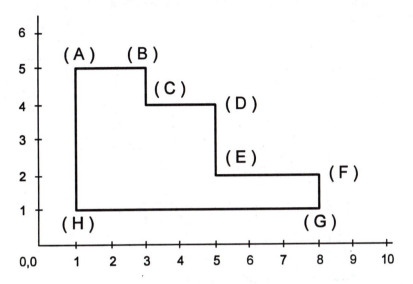

Reprinted with permission from McGrew,
Exploring the Power of AutoCAD, copyright 1990 by Delmar Publishers.

25. Determine the perimeter of the CAD drawing if the grid spacing is ⁷⁄₁₆". _____

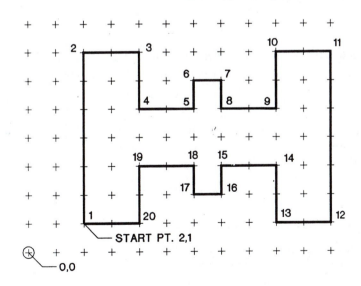

Reprinted with permission from Stellman, Krishnan, and Rhea,
Harnessing AutoCAD, copyright 1993 by Delmar Publishers.

26. Find the total cost of a box of twelve 5 $\frac{1}{4}$ " floppy disks if one disk costs 87 $\frac{1}{2}$ cents.

_____

27. What would be the new radius sizes on the CAD drawing if the gasket is enlarged by a factor of 5?

A _____

B _____

C _____

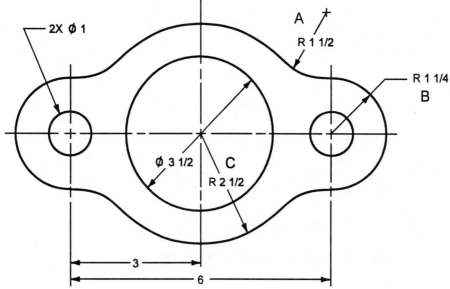

2X Ø 1

A

R 1 1/2

R 1 1/4

B

Ø 3 1/2

C

R 2 1/2

3

6

28. The CAD drawing below will be inserted into another project as a symbol. Determine the new fractional sizes if it is increased by a factor of 3 when inserted.

A _____

B _____

C _____

A

7/8

2

3/4 B

4 1/4

C

29. Using the CAD drawing shown, find new dimensions **A** and **B** if it is scaled up by a factor of 7.

A _____

B _____

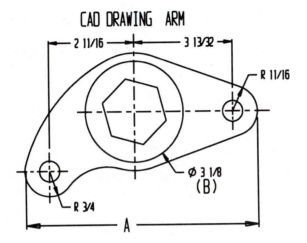

CAD DRAWING ARM

2 11/16   3 13/32

R 11/16

Ø 3 1/8
(B)

R 3/4

A

30. Find the new overall sizes for the parts illustrated in the CAD drawing. A factor of 8 will be used to enlarge both parts.

Height _____

Width _____

Depth _____

Rod diameter _____

Rod length _____

Ø7/16 – 2 HOLES

1/2   7/8   1 1/4

NOTE:
MAT'L. = BRASS
2 REQUIRED
ROD LENGTH = 3 1/4

45°

1/2   3 1/2

4 1/2

1   3/4   1   3/4

5/8

5/8–10ACME–2

1/2

2 1/2

1 13/16   1 1/8

1/2   11/16   1/2

1 9/16   1 3/8

Unit 9 DIVISION OF COMMON FRACTIONS

BASIC PRINCIPLES OF DIVISION OF COMMON FRACTIONS

When dividing a common fraction, invert the divisor and proceed as in multiplication. Remember to reduce answers to the lowest terms.

Example: Divide $\frac{1}{4}$ by $\frac{3}{8}$.

$$\frac{1}{4} \times \frac{8}{3} = \frac{1 \times 8}{4 \times 3} = \frac{8}{12} = \frac{2}{3}$$

When dividing fractions and whole numbers or mixed numbers, the entire problem must be changed to fractions.

Example: Divide 3 by $\frac{1}{5}$.

$$3 \div \frac{1}{5} = \frac{3}{1} \div \frac{1}{5} = \frac{3}{1} \times \frac{5}{1} = \frac{3 \times 5}{1 \times 1} = \frac{15}{1} = 15$$

Example: Divide $2\frac{1}{8}$ by $1\frac{1}{2}$.

$$2\frac{1}{8} \div 1\frac{1}{2} = \frac{17}{8} \div \frac{3}{2} = \frac{17}{8} \times \frac{2}{3} = \frac{17 \times 2}{8 \times 3} = \frac{34}{24} = \frac{17}{12} = 1\frac{5}{12}$$

⊞ 17 ⓐ⅟c 8 (÷) 3 ⓐ⅟c 2 (=) 1⁵⁄₁₂

PRACTICAL PROBLEMS

Note: The units of measure are divided in the division of fractions.

1. $\frac{7}{8}$ inch ÷ $\frac{7}{32}$ inch = _____

2. $5\frac{3}{8}$ lb. ÷ $3\frac{3}{4}$ lb. = _____

3. 17 yd. ÷ $\frac{5}{6}$ yd. = _____

4. $\frac{15}{16}$ inch ÷ 8 = _____

5. $\frac{13}{16}$ inch ÷ $\frac{1}{8}$ = _____

6. $5\frac{3}{4}$ yd. ÷ $\frac{1}{3}$ = _____

7. $\frac{3}{7}$ ÷ $4\frac{2}{3}$ = _____

70

8. $2\frac{3}{16}$ inch ÷ $1\frac{1}{2}$ inch = _____

9. $2\frac{7}{10}$ cm ÷ $6\frac{3}{10}$ cm = _____

10. $4\frac{1}{2}$ ÷ $\frac{7}{8}$ ÷ $\frac{1}{4}$ = _____

11. A regular octagon has eight equal sides. The perimeter of the octagon is $11\frac{3}{8}$ inches. Find, in inches, the length of each side. _____

12. A piece of drafting lead is $37\frac{5}{8}$ inches long. How many $5\frac{3}{8}$ inches drafting leads can be cut from it? (No allowance is made for waste.) _____

13. All steps on this template are equal. Find, in inches, dimension **A**. _____

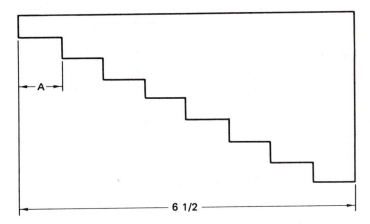

14. How many pieces of metal, each $\frac{27}{64}$" in length, can be made from a strip 54" in length? Assume there is no waste in cutting. _____

15. A CAD drafter completes a certain job in $\frac{2}{3}$ hour. How many jobs of the same kind can be completed in 8 hours? _____

16. All segments of this sheet metal layout of the truncated cylinder are equal in width. Find, in inches, dimension **A**.

_____

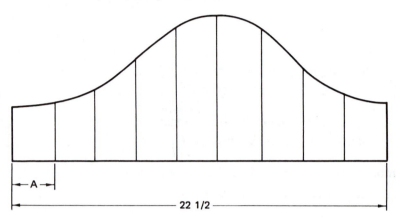

17. A piece of ⅛" diameter brass rod is 30" long. How many ³⁄₁₆" pins can be sheared from the rod?

_____

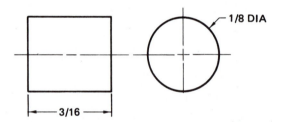

18. Screw threads are classified by the number of threads per inch. The pitch of a screw thread is expressed as a fraction. The numerator is one. The denominator is the number of threads per inch. This bolt has a pitch of ¹⁄₁₆. Determine the number of threads on the bolt.

_____

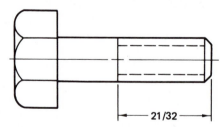

19. A stack of ⁵⁄₁₆" thick erasers is 4¹¹⁄₁₆" high. How many erasers are in the stack?

_____

20. The perimeter of a regular hexagon is 19½". Find, in inches, the length of each side.

_____

21. Equal divisions are drawn on this displacement diagram for a cam layout. Determine dimension **A** in inches. _____

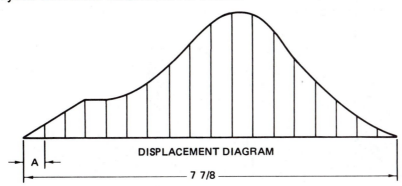

DISPLACEMENT DIAGRAM

A |←

7 7/8

22. A strip of $\frac{7}{16}$-inch metal is 161 inches long. How many pieces, each $\frac{7}{8}$ inch long, can be stamped from it? _____

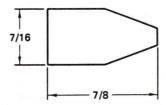

7/16

7/8

23. On a drawing, $\frac{1}{8}$ inch equals 1 inch.

a. How many inches are represented by a line drawn $3\frac{3}{16}$ inches long? _____

b. How many inches are represented by a line drawn $3\frac{13}{32}$ inches long? _____

24. The isometric CAD drawing shown must be scaled down by a factor of 3 to make room for additional views. The present height is $2\frac{1}{4}$", width is $10\frac{1}{8}$", and the depth measures $8\frac{1}{4}$". Find the new spatial dimensions.

Height _____

Width _____

Depth _____

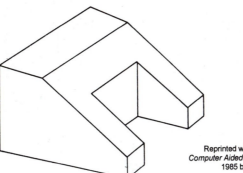

25. The CAD drawing shown using grid points has a width of 8¾" and height of 6⅛". Determine the unit of spacing between grid points. Disregard the angular measurements.

_____

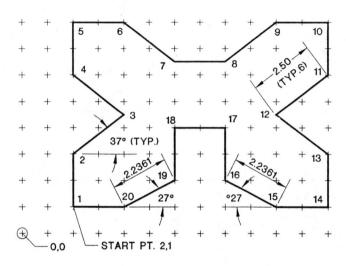

26. The DIVIDE command is used to divide a line 9⅜" long into 15 equal divisions. How long is each new division?

_____

27. The CAD drawing shown below is to be scaled down by a factor of 4. Find the new sizes of **A** and **B** once the scaling operation has been performed.

A _____

B _____

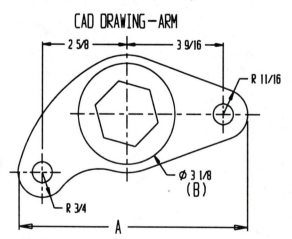

CAD DRAWING—ARM

28. A CAD drafter reduces the arm by a factor of 8. What will be the new sizes for the diameters shown?

A _____

B _____

C _____

D _____

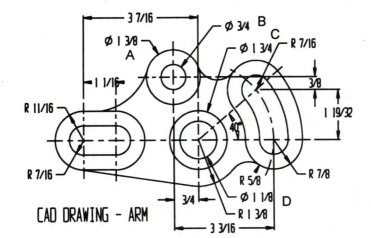

CAD DRAWING - ARM

29. Using the CAD drawing below, determine the new sizes for dimensions **A** and **B** if the drawing is scaled down by a factor of 5.

A _____

B _____

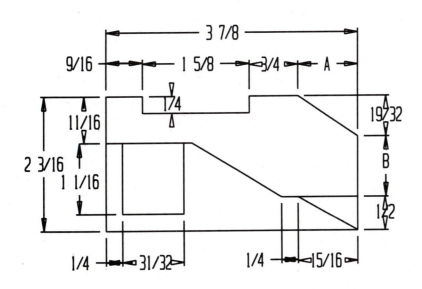

30.   A CAD operator inserts the CAD drawing below as a symbol. During the
insertion process, the X-axis (**A**) is reduced using a factor of 3 while the
Y-axis (**B**) is reduced using a factor of 4. Find the new values for
dimensions **A** and **B**.

A _____

B _____

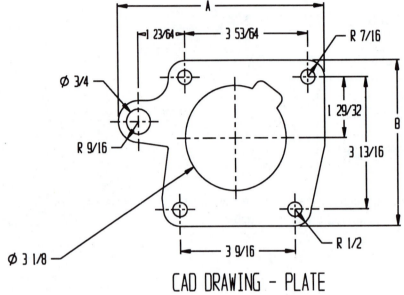

CAD DRAWING - PLATE

Unit 10 COMBINED OPERATIONS WITH COMMON FRACTIONS

BASIC PRINCIPLES OF COMBINED OPERATIONS

This unit presents practical problems involving combined operations of addition, subtraction, multiplication, and division with common fractions.

PRACTICAL PROBLEMS

1. $(\frac{3}{8}" + 1\frac{15}{16}") \times \frac{7}{8} =$ _____

2. $(\frac{5}{6} \text{ hr.} - \frac{1}{3} \text{ hr.}) \times 8 =$ _____

3. $3\frac{1}{4}$ pounds $\div \frac{3}{4}$ pounds plus 6 = _____

4. $36\frac{1}{2}$ meters minus $17\frac{3}{4}$ meters plus $11\frac{1}{4}$ meters = _____

5. $3\frac{3}{4} \times 3\frac{1}{3} \div 2\frac{1}{2} =$ _____

6. Determine, in inches, dimensions **A**, **B**, and **C** on this gasket.

    A _____

    B _____

    C _____

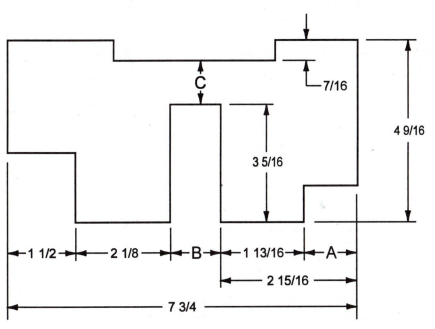

7.   Which group is thicker: 10 pieces of $\frac{1}{4}$" brass or 6 pieces of $\frac{7}{16}$" aluminum?

8.   A block of steel is $1\frac{15}{16}$" thick. A rough cut of $\frac{3}{16}$" and a finish cut of $\frac{7}{64}$" are taken. What is the remaining thickness?

9.   A civil drafter charges these times to a certain project: 8 hours, $6\frac{1}{2}$ hours, $7\frac{1}{4}$ hours, $5\frac{2}{3}$ hours, and $8\frac{1}{3}$ hours. How many more working hours are needed to total 40?

10. Find, in inches, the inside diameter (dimension **D**) of this cross section of pipe.

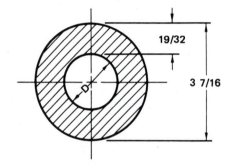

11. Ten erasing shields can be stamped in $7\frac{1}{2}$ seconds. At the same rate, how many can be stamped in one minute?

12. On this stretchout, the widths of all sections are equal. Dimension **B** is one-half of dimension **A**. Find, in inches, dimensions **A** and **B**.

A _____

B _____

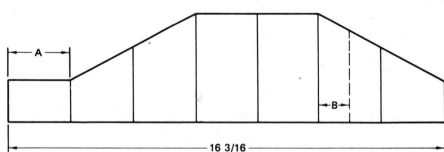

13. A stock rack contains four pieces of drill rod having lengths of $4\frac{1}{2}$", $3\frac{3}{16}$", $7\frac{1}{8}$", and $5\frac{13}{16}$". Determine the length of another piece to make a total length of $26\frac{23}{32}$".

14. On this strap dimension **A** is $\frac{7}{16}$" and dimension **B** is $1\frac{3}{32}$". Find, in inches, dimension **C**. _____

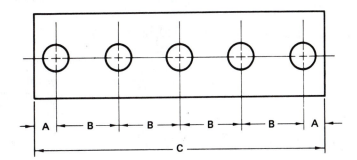

15. On a scale drawing, $\frac{1}{4}$" represents 1'. A line is $7\frac{3}{4}$" long. How many feet does it represent? _____

16. The sides of the pentagonal cutout on the shim are $7\frac{1}{4}$". The sides of the octagonal bar stock are $4\frac{3}{8}$". Which polygonal shape has the larger perimeter? _____

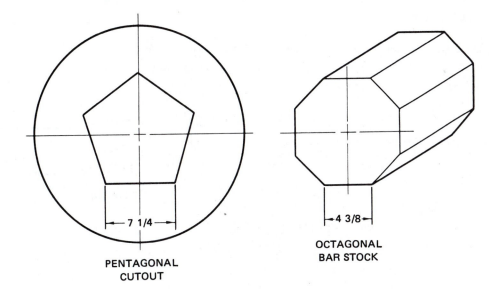

7 1/4

PENTAGONAL
CUTOUT

4 3/8

OCTAGONAL
BAR STOCK

17. Cylindrical handles are cut from handle stock. The lengths available are $38\frac{1}{2}$", $49\frac{1}{2}$", $93\frac{1}{2}$", $60\frac{1}{2}$". How many $5\frac{1}{2}$" handles can be made? _____

18. The threaded length of a threaded fastener under 6 inches in length is usually 2 times the fastener diameter plus $\frac{1}{4}$". Find the threaded length of a 2 $\frac{1}{2}$" bolt with $\frac{5}{8}$" diameter.

_____

19. The holes on this template are equally spaced. Find, in inches, the overall length.

_____

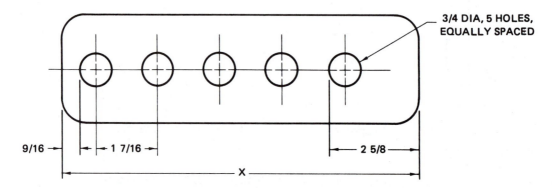

3/4 DIA, 5 HOLES, EQUALLY SPACED

9/16   1 7/16   2 5/8   X

20. Seven holes are equally spaced with $\frac{13}{16}$" between the centers of the holes. Determine the distance from the center of the first hole to the center of the fifth hole.

_____

21. Using the CAD drawing and the dimensions provided, find the perimeters of the cabin and the deck. What is the difference of the perimeters? State your answers in feet and inches.

Cabin perimeter _____

Deck perimeter _____

Difference of perimeters _____

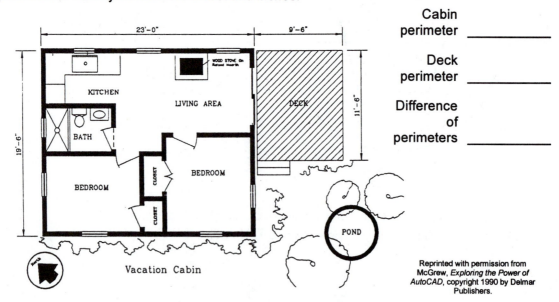

Vacation Cabin

Reprinted with permission from McGrew, *Exploring the Power of AutoCAD*, copyright 1990 by Delmar Publishers.

22. The CAD drawing to the right has been dimensioned to indicate its new sizes. The drawing was scaled down by a factor of 5. Find the original height and width sizes. A copy was placed to the left using a scale factor of 9 times the hole diameter. What is the displacement distance of the copied view (point **A** to point **B**)?

Height _____

Width _____

Distance _____

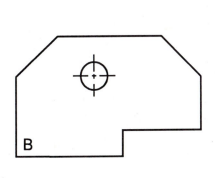

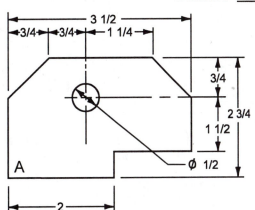

Reprinted with permission from McGrew,
Exploring the Power of AutoCAD, copyright 1990 by Delmar Publishers.

23. Using the CAD drawing presented, find the overall height and width dimensions as well as the sizes of features **A** and **B**.

Height _____

Width _____

A _____

B _____

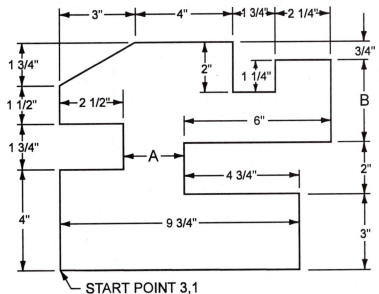

Reprinted with permission from Stellman,
Krishnan, and Rhea, *Harnessing AutoCAD,*
copyright 1993 by Delmar Publishers.

24. Find dimension **A** on the CAD drawing. Determine the radii sizes if the drawing is scaled up by a factor of 5.

A _____

New radius **A** _____

New radius $^{11}\!/_{16}$ _____

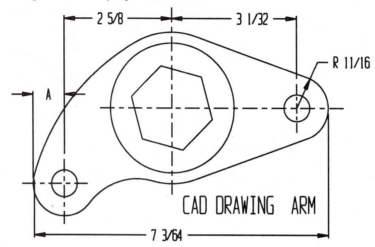

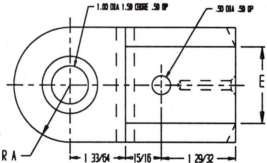

2 5/8   3 1/32

R 11/16

A

CAD DRAWING   ARM

7 3/64

25. Using the CAD drawing below, determine the sizes for dimensions **A–E**.

A _____

B _____

C _____

D _____

E _____

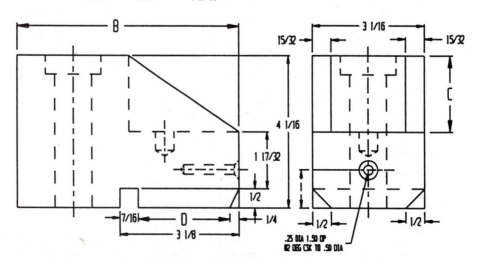

1.00 DIA 1.50 CBORE .50 DP   .50 DIA .50 DP

E

R A   1 33/64   15/16   1 29/32

B

15/32   3 1/16   15/32

4 1/16

C

1 17/32

1/2

7/16   D   1/4

3 1/8

.25 DIA 1.50 DP
82 DEG CSK TO .50 DIA

1/2   1/2

CAD DRAWING - LOCATING FIXTURE

26. In the diagram below, find dimensions **D** and **F** if **A** is $\frac{9}{16}$", **B** is $1\frac{3}{8}$", **C** is $2\frac{29}{32}$", and **E** is $3\frac{7}{64}$". Also find dimension **Y** if **X** is $\frac{33}{64}$" and **W** is $2\frac{9}{64}$".

D _____

F _____

Y _____

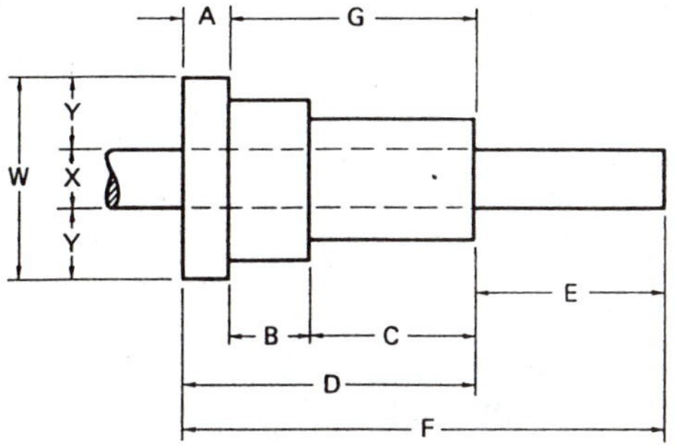

Reprinted with permission from Herman and Gerard,
Practical Problems in Mathematics for Electricians, copyright 1996 by Delmar Publishers.

27. Find the dimensions **A–D** using the CAD drawing and dimensions provided. State your answers in feet and inches. Also find the inside perimeter of the living room and dining area. Disregard the two doors, archway, and fireplace.

A _____

B _____

C _____

D _____

Perimeter _____

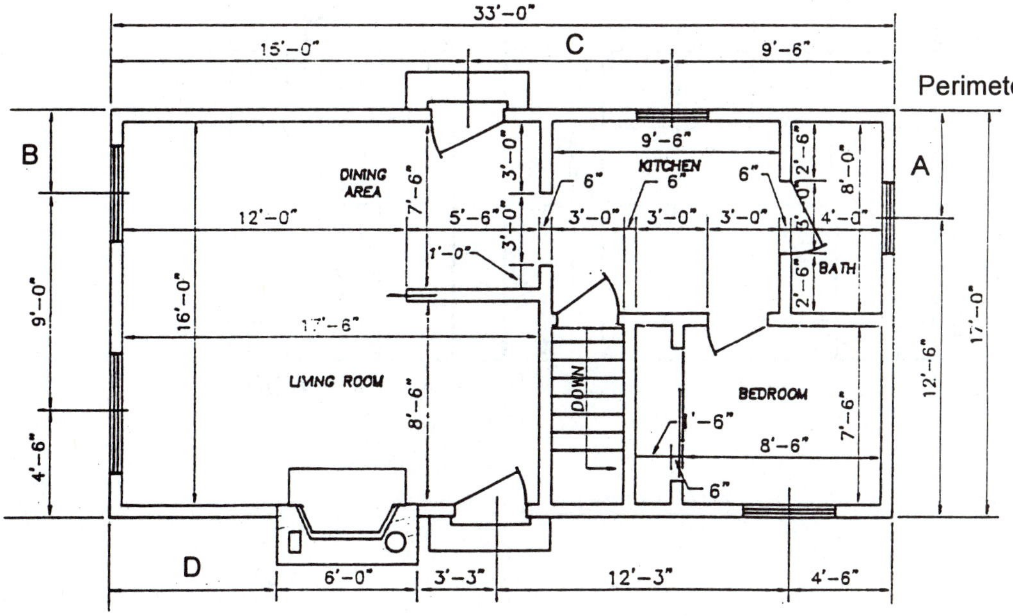

28. Using the CAD drawing provided, find the sizes for dimensions **A**, **B**, and **C**. What would be the size of the large radius if the top view were increased by a factor of 4?

A _____

B _____

C _____

Radius _____

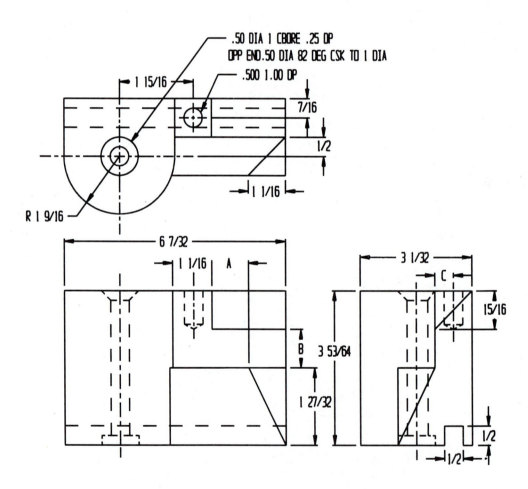

29. Find the size of dimension **A** using the CAD drawing presented. Also determine the size of the two diameters if the object is to be scaled up using a factor of 5.

A _____

Diameter 1 _____

Diameter 2 _____

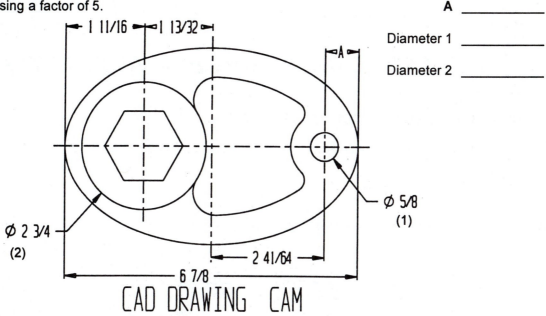

CAD DRAWING CAM

30. Using the CAD drawing illustrated, find the difference between the outer perimeter of the shim and the sum of the perimeters of the inner features. The shim is symmetrical.

Inner perimeters (sum) _____

Outer perimeter _____

Difference _____

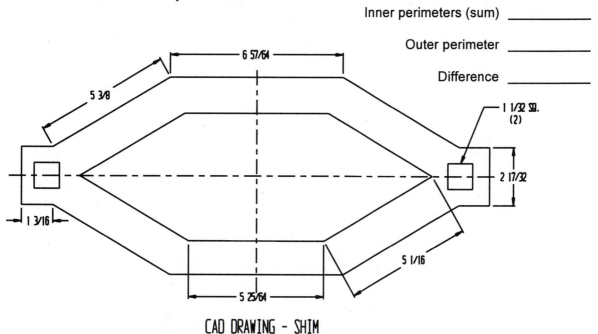

CAD DRAWING - SHIM

Decimal Fractions

Unit 11 ADDITION OF DECIMAL FRACTIONS

BASIC PRINCIPLES OF ADDITION OF DECIMAL FRACTIONS

Decimal fractions are fractions with denominators which are powers of ten.

Example: $\dfrac{5}{10}$, $\dfrac{9}{100}$, $\dfrac{17}{1000}$

When these fractions are written in decimal form, they are referred to as decimal fractions or decimals.

Example: $\dfrac{5}{10}$ = 0.5, $\dfrac{9}{100}$ = 0.09, $\dfrac{17}{1000}$ = 0.017

To add decimals, the decimal points must be located directly under each other, then added to find the sum.

Example: *12 1*
            0.7
            2.6
         141.756
       +36.08
       ──────
       181.136

   0.7 ⊞ 2.6 ⊞ 141.756 ⊞ 36.08 ⊟ 181.136

PRACTICAL PROBLEMS

Add the following quantities.

1.	0.75	2.	0.245	3.	0.5018	4.	0.987
	+ 0.137		+ 0.767		+ 0.375		+ 0.33

5. 0.7325 + 0.112 + 0.416 =   _____

6. 2.1307 + 3.718 + 4.66 =   _____

7. 21.601 + 7.009 + 16.736 =   _____

8. 0.007 + 4.25 + 14.379 + 8.3 =   _____

9. What is the overall length, in millimeters, of this link?   _____

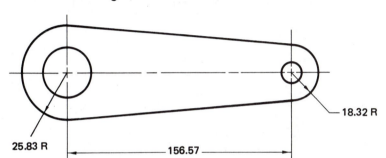

18.32 R

25.83 R   156.57

10. Many working drawings are toleranced dimensions. A *tolerance* is the difference that is allowed in the size of a part. The dimensions on this shim show the basic size and the tolerance. Find, in inches, the *largest* length and width permitted.

A _____

B _____

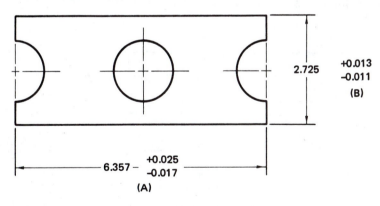

2.725   +0.013 / −0.011   (B)

6.357 − +0.025 / −0.017

(A)

11. Find the total cost of the following drafting instruments: 6" bow compass, $7.85; mechanical drafter's scale, $3.75; drafter's mechanical pencil, $1.65; and an 8" adjustable triangle, $8.39.

_____

12. The thicknesses of six machine parts are measured with a micrometer. The measurements are 0.065", 0.145", 0.7146", 1.606", 0.329", and 3.125". What is the total thickness, in inches, of the parts?

_____

13. In one week a civil drafter works 8.75 hours, 10.25 hours, 6.5 hours, 8 hours, and 7.25 hours. How many hours does the drafter work? _____

14. Find, in inches, the perimeter of this gasket. _____

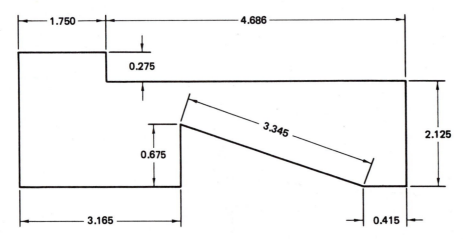

15. A drafter's weekly paycheck has the following deductions: $8.10 medical insurance, $56.13 federal income tax, $12.17 F.I.C.A., $4.10 state income tax, and $2.35 union dues. Compute the total deductions. _____

16. Calculate, in millimeters, the overall length of this sliding lever. _____

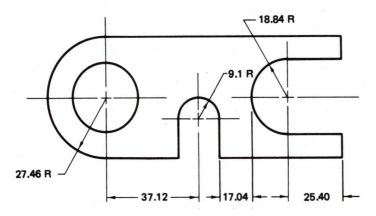

17. On a center line, a mechanical drafter measures off 3.75 centimeters, 6.25 centimeters, 0.60 centimeters, 7.37 centimeters, and 5.60 centimeters. What is the overall length of the line in centimeters? _____

18. Find, in inches, dimensions **A** and **B** on this strap.

A _____

B _____

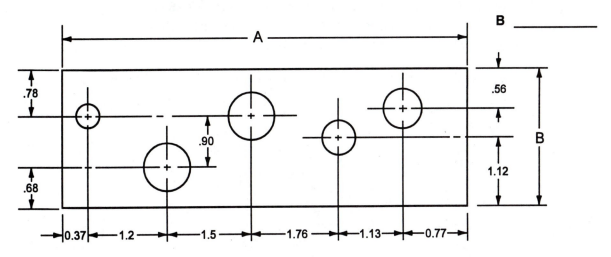

19. In one year a drafter receives $37.50, $18.75, $21.60, $45.80, $27.33, and $12.88 in overtime pay. What is the total overtime pay? _____

20. Determine, in inches, dimension **A** on this step block. _____

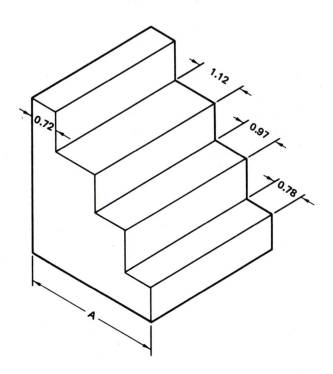

21. A CAD drafter has devoted several weeks to a structural steel project. The hours spent on various aspects of the job are 40.05 hrs., 37.25 hrs., 146.75 hrs., .40 hrs., 92.45 hrs., and 112.15 hrs. Find the total number of hours the CAD drafter has worked on this project. _____

22. Determine the height and width of the CAD drawing provided. Height _____

    Width _____

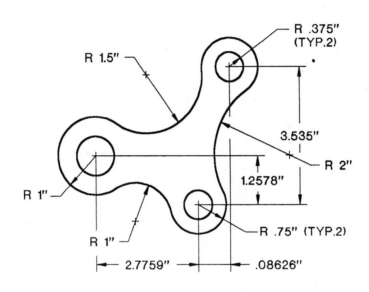

Reprinted with permission from Stellman, Krishnan, and Rhea,
Harnessing AutoCAD, copyright 1993 by Delmar Publishers.

23. A CAD drafter purchases a CAD station for use at home. The computer and software costs $5,675.00. Later a plotter is added at a cost of $965.75, a printer at $449.95, a FAX modem at $127.50, and a CD-ROM at $368.50. What is the total cost of this investment? _____

24. Using the CAD drawing of the plot plan, find its perimeter in feet. _____

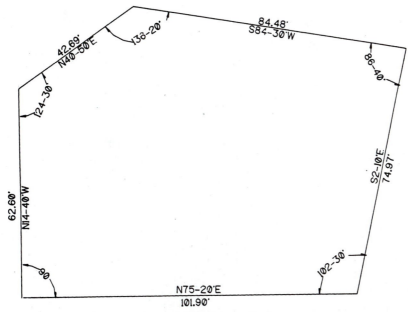

Reprinted with permission from Zandi,
Computer Aided Drafting and Design, copyright 1985 by Delmar Publishers.

25. Determine the overall length of the CAD drawing shown. State your answer in inches. _____

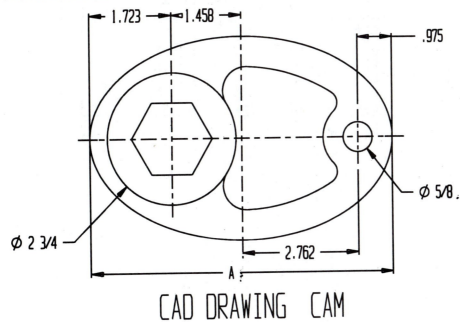

CAD DRAWING CAM

26.   Calculate dimensions **A, B,** and **C** using the CAD drawing presented.

A _____

B _____

C _____

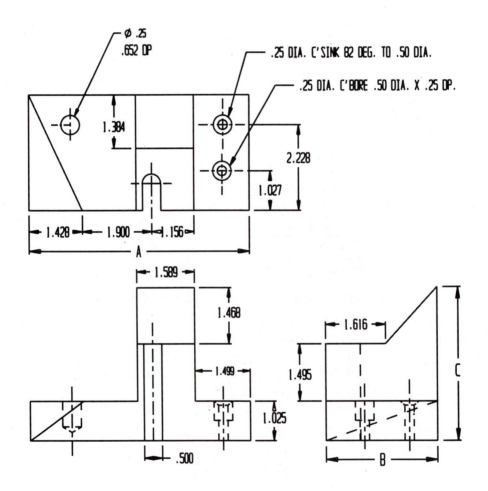

27. Using the CAD drawing of the arm, find, in inches, dimension **A**. _____

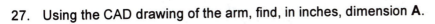
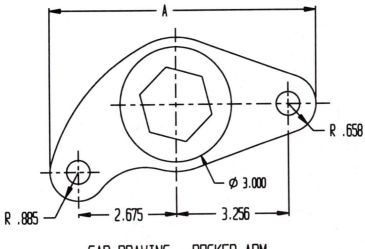

CAD DRAWING - ROCKER ARM

28. Find dimensions **A**, **B**, and **C** using the CAD drawing shown. Figures are
     in inches.

A _____

B _____

C _____

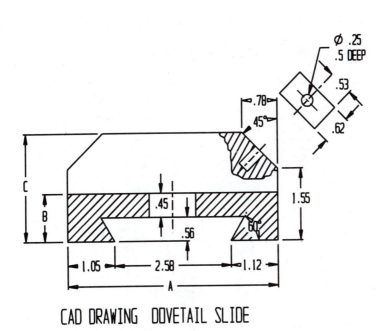

CAD DRAWING DOVETAIL SLIDE

29. Using the CAD drawing provided, find the overall height and width in inches.

Height _____

Width _____

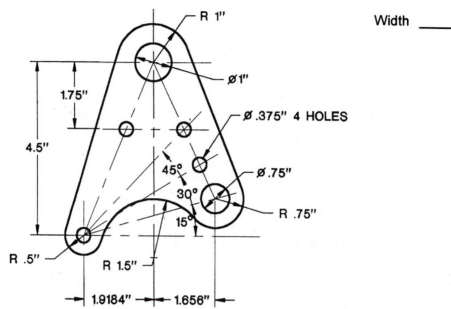

R 1"

Ø1"

1.75"

Ø.375" 4 HOLES

4.5"

45°

Ø.75"

30°

R .75"

15°

R .5"

R 1.5"

1.9184"   1.656"

Reprinted with permission from Stellman, Krishnan, and Rhea, *Harnessing AutoCAD,* copyright 1993 by Delmar Publishers.

30. Determine dimensions **A** and **B** in inches using the CAD drawing provided.

A _____

B _____

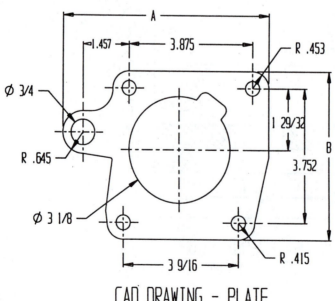

A

1.457   3.875   R .453

Ø 3/4

R .645

1 29/32

B

3.752

Ø 3 1/8

3 9/16   R .415

CAD DRAWING - PLATE

Unit 12 SUBTRACTION OF DECIMAL FRACTIONS

BASIC PRINCIPLES OF SUBTRACTION OF DECIMAL FRACTIONS

When subtracting decimal fractions, place the smaller number under the larger number. Make sure the decimal points are aligned properly.

Example: 19.643 – 8.132

$$
\begin{array}{r}
19.643 \\
-\ 8.132 \\
\hline
11.511
\end{array}
$$

  19.643 ⊖ 8.132 ⊜ 11.511

PRACTICAL PROBLEMS

Subtract the following quantities.

1.	0.87	2.	0.342	3.	1.721	4.	8.375
	– 0.28		– 0.063		– 0.659		– 3.683

5. 18.6 – 10.89 =   _____

6. 21.634 – 9.357 =   _____

7. 118.717 – 77.9 =   _____

8. 675.6 – 486.857 =   _____

9. Find, in millimeters, the length of the threaded portion of this round head
    cap screw.   _____

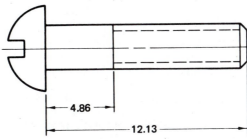

4.86

12.13

10. The outside diameter of a shaft tapers from 2.8750" to 1.2188". What is the difference in these two diameters?

_____

11. A block of steel is 25.40 mm thick. One cut of 3.25 mm and a second cut of 4.17 mm are taken. What is the remaining thickness, in millimeters, of the block?

_____

12. Two pins are to be located in a steel base. The distance between the centers of the pins must be determined. Find, in inches, dimension **A**.

_____

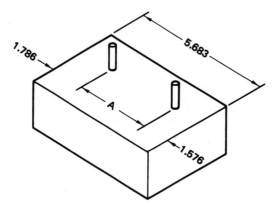

13. A civil drafter has a $425 budget for equipment. A new drafting machine for $233, machine scales for $13.58, and an electronic calculator for $117.86 are purchased. How much money is left in the budget?

_____

14. The length of the front view is 257.37 mm. The width of the right profile view is 147.82 mm. Find, in millimeters, the difference between these measurements.

_____

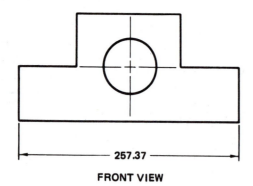

FRONT VIEW

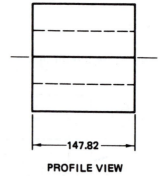

PROFILE VIEW

15. Two reams of paper weigh 3.87 pounds and 14.35 pounds, respectively. Find, in pounds, the difference in the weights.

_____

16. An architectural drafter's bill for new equipment is $167.74. The discount is $16.75. How much does the drafter actually pay for the new equipment? _____

17. A steel block is 81.315 centimeters long. A machinist cuts off 6.250 cm. What is the length of the block, in centimeters, after machining? _____

18. What is the *smallest* permissible size, in inches, for each dimension on this set-up block?

A _____

B _____

C _____

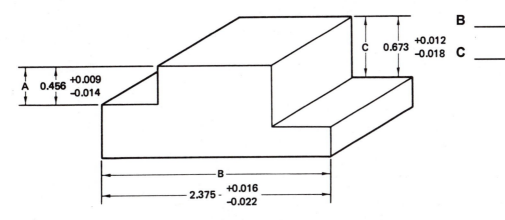

19. The outside diameter *(OD)* of a drilled hole is 0.7812". The hole is made larger with a reamer having an outside diameter *(OD)* of 0.7969". Determine, in inches, the increase in diameter after the reaming operation. _____

20. Find, in centimeters, dimensions **A** and **B** of this shim.

A _____

B _____

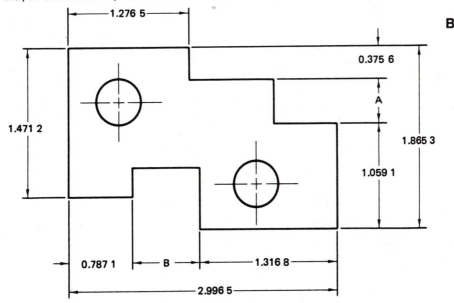

21. A clearance fit of .012 is needed between a shaft and a bushing to work properly. The bushing diameter *(ID)* measures 1.475 inches. What must the shaft diameter *(OD)* measure to obtain this clearance?

_____

22. Find dimension **A** in inches using the CAD drawing illustrated.

_____

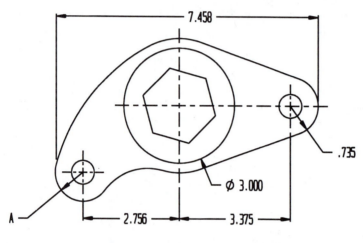

CAD DRAWING - ROCKER ARM

23. A CAD operator makes the following purchases: $849.50 for CAD software, $27.75 for disk management software, $1,286.00 for a memory upgrade kit, and $535.95 for a paint program. A $37.50 cash rebate was given at the time of purchase. How much did the CAD operator pay for these products?

_____

24. Find dimensions **A, B,** and **C** in inches using the CAD drawing shown.

A _____

B _____

C _____

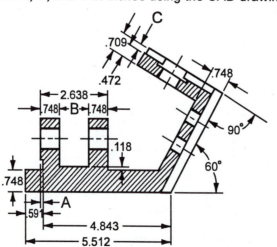

25. The CAD drawing presented has a perimeter of 22.3297". Find the missing dimension for the length of line **AB**.

_____

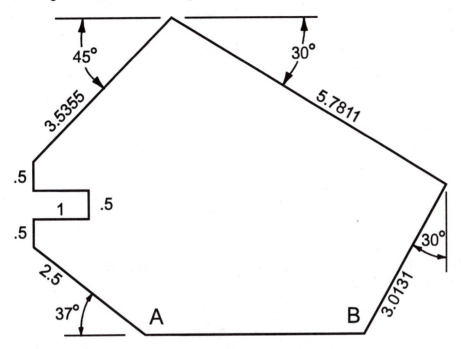

Reprinted with permission from McGrew,
Exploring the Power of AutoCAD, copyright 1990 by Delmar Publishers.

26. Using the CAD drawing below, find the difference between the outside perimeter and the sum of the perimeters of the inside features.

_____

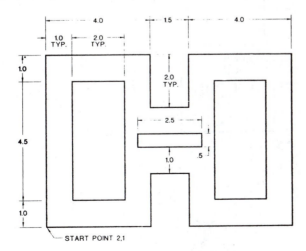

Reprinted with permission from Stellman, Krishnan, and Rhea,
Harnessing AutoCAD, copyright 1993 by Delmar Publishers.

27. Determine dimension **A** in inches using the CAD drawing shown.

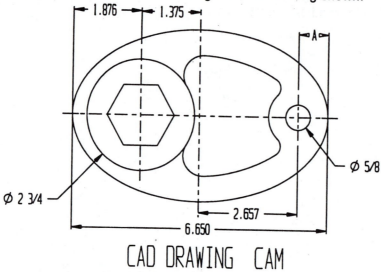

CAD DRAWING CAM

28. The perimeter of the CAD drawing is 2812.9'. Find the missing dimension between points **E** and **F**.

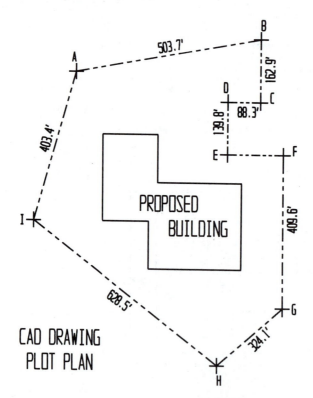

CAD DRAWING
PLOT PLAN

29. Using the CAD drawing given, find in inches the difference between the
   outer perimeter and the sum of the perimeters of the inner features. The
   shim is symmetrical.

   Outer _____

   Sum—Inner _____

   Difference _____

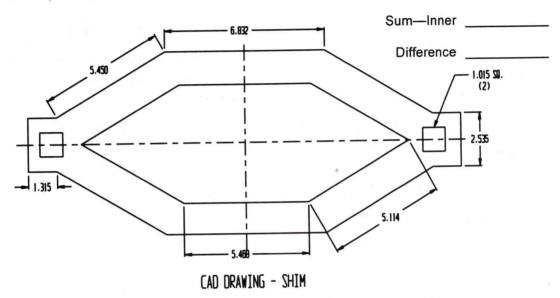

CAD DRAWING – SHIM

30. Determine the dimensions **A–E** in centimeters on the CAD drawing of
   the retainer.

   A _____

   B _____

   C _____

   D _____

   E _____

CAD DRAWING – RETAINER

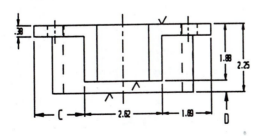

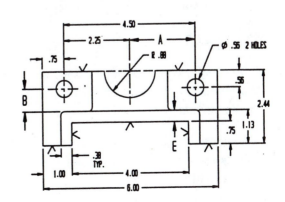

Unit 13 MULTIPLICATION OF DECIMAL FRACTIONS

BASIC PRINCIPLES OF MULTIPLICATION OF DECIMAL FRACTIONS

Decimal fractions are multiplied in the same manner as whole numbers. The number of decimal places to the right of the decimal point in both numbers being multiplied are added. Count this number in from the right and place the decimal point in the answer.

Examples:

12.15	(two decimal places)
× 2.3	(one decimal place)
3645	
2430	
27.945	(three decimal places)

.003
× .017
021
003
.000051

  12.15 $\boxed{X}$ 2.3 $\boxed{=}$ 27.945

PRACTICAL PROBLEMS

Multiply the following quantities.

1.
```
   0.83
 × 0.06
```

2.
```
   0.78
 × 0.23
```

3.
```
   0.465
 × 0.87
```

4.
```
   1.714
 × 0.32
```

5. 13.64 × 0.27 =   _____

6. 27.321 × 0.65 =   _____

7. 32.18 × 0.69 =   _____

8. 43.76 × 0.58 =   _____

9. A bolt has a diameter of 1.375 inches. The threaded length of a bolt is two times the diameter plus 0.250 inch. What is the threaded length of the bolt in inches?   _____

10. A designer knows that for each complete turn a wheel rolls a distance of 3.14 times its diameter. How far does a 12.6" diameter wheel roll in 3 complete turns?   _____

11. A shaft is tapered 0.375" per inch of its length. How many inches is this shaft tapered? _____

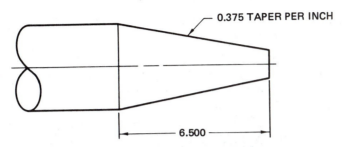

0.375 TAPER PER INCH

6.500

12. A certain type of mylar film costs 12.5 cents a sheet. What is the cost of 144 sheets? _____

13. A mechanical drafter's plastic scale weighs 0.632 pounds. What is the weight of 48 scales? _____

14. A spacing collar is 0.375" wide. How many inches will 17 spacing collars take up on a shaft? _____

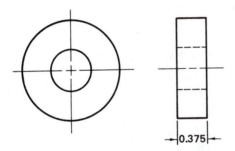

0.375

15. A CAD drafter's pay is increased $37.68 per month. What is the total raise for one year? _____

16. All sections on this gauge are equal in width. Find the overall length in decimeters. _____

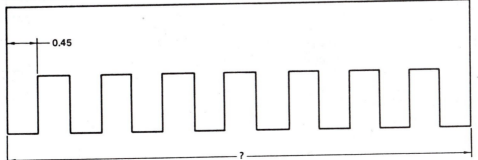

0.45

?

17. An isometric ellipse is 1.225 times larger than the circle it represents. A circle is 0.750" in diameter. What is the size of the ellipse in inches?   _____

18. There are 17 equally spaced holes on this plate. Find, in inches, the distance between the centers of holes **A** and **B**.   _____

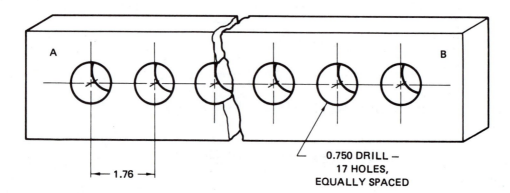

0.750 DRILL –
17 HOLES,
EQUALLY SPACED

1.76

19. A microfilmed drawing is enlarged 2.75 times. A line on the drawing is 5.25 cm long. What is the length, in centimeters, of the enlarged line?   _____

20. This round head cap screw has a pitch of $\frac{1}{12}$ or 12 threads per inch. How many threads are contained in a rod that is 9.5 inches long and has the same thread pitch as the cap screw shown?   _____

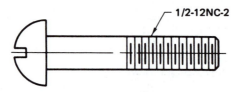

1/2-12NC-2

21. The circumference of a circle equals the diameter × 3.1416. What is the circumference, to the nearest hundredth, for a wheel that has a diameter of 8.5 inches?   _____

22. A CAD drafter travels 13.30 miles to work each day. Find the total round trip mileage if the drafter worked 165.50 days.   _____

23. Determine the height, width, and depth in centimeters once the CAD drawing is scaled up by a factor of 8.

Height _____

Width _____

Depth _____

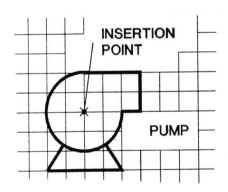

.56

1.07

3.16

1.58

CAD DRAWING - ISOMETRIC

24. The cost of a three-button mouse is $79.65. What is the cost of six mice of this model?

_____

25. The pump symbol is to be inserted in a CAD drawing at 3 times its present size. Find the new size if the grid size is .125.

Height _____

Width _____

INSERTION POINT

PUMP

26. In the CAD drawing presented, the radii must be enlarged by a factor of 4 to receive another part. Also, the larger holes must be increased by a factor of 2.75 and the small hole must be increased by a factor of 2.25. Determine the size in inches of these new enlargements.

Radii _____

Large holes _____

Small hole _____

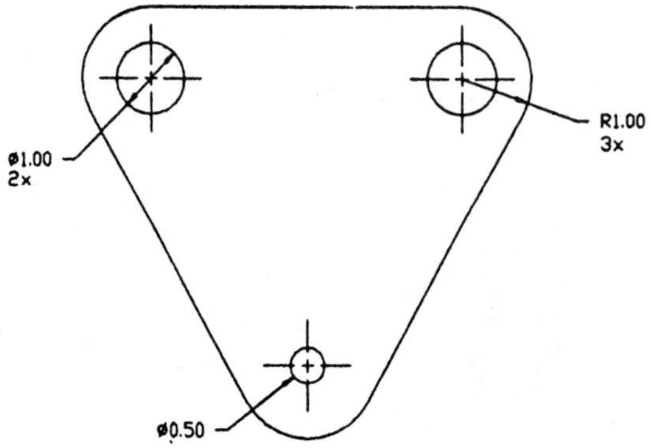

R1.00
3x

⌀1.00
2x

⌀0.50

Reprinted with permission from Resetarits and Bertolini,
Using CADKEY Light, copyright 1992 by Delmar Publishers.

27. The symbol shown below will be inserted into a CAD drawing 7 times larger than its existing size. If the current grid spacing is .375, find the new height and width of the symbol once it is inserted.

Height _____

Width _____

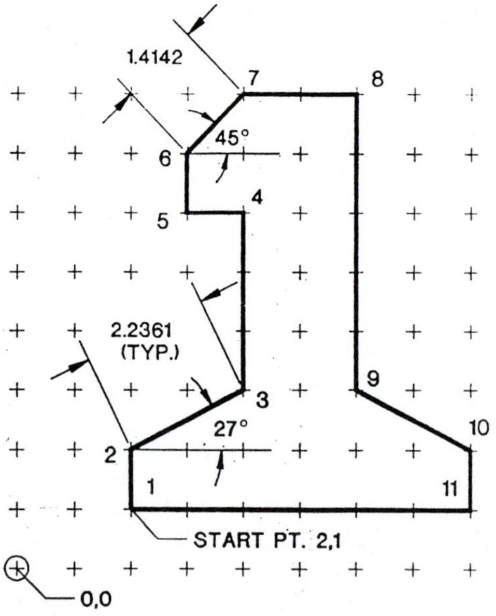

1.4142

45°

2.2361
(TYP.)

27°

START PT. 2,1

0,0

Reprinted with permission from Stellman, Krishnan, and Rhea,
Harnessing AutoCAD, copyright 1993 by Delmar Publishers.

28. A 5 ¼" high density diskette costs $1.79. Find the cost of a dozen diskettes. What would be the cost of a gross of diskettes?

Dozen _____

Gross _____

29. The exchanger symbol used in piping is shown using a .25 grid. The symbol is inserted using a 1.75 scaling factor. Find the enlarged sizes for the height and width.

Height _____

Width _____

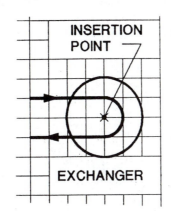

INSERTION POINT

EXCHANGER

Reprinted with permission from Stellman, Krishnan, and Rhea,
Harnessing AutoCAD, copyright 1993 by Delmar Publishers.

30. The CAD drawing illustrated will be scaled up by a factor of 9. Find the new dimensions for length and width.

Length _____

Width _____

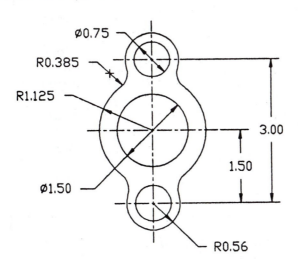

Ø0.75

R0.385

R1.125

Ø1.50

3.00

1.50

R0.56

Reprinted with permission from Resetarits and Bertolini,
Using CADKEY Light, copyright 1992 by Delmar Publishers.

Unit 14 DIVISION OF DECIMAL FRACTIONS

BASIC PRINCIPLES OF DIVISION OF DECIMAL FRACTIONS

When dividing decimal fractions, the divisor must be a whole number, not a fraction. To make the divisor a whole number, the decimal point is moved all the way to the right of the number. The decimal point of the dividend is then moved the same number of places to the right. The decimal point of the quotient or answer is located above the decimal point of the dividend.

Example:

$$
\begin{array}{r}
1\,7.6 \\
5.9\,)\overline{103.8.4} \\
\underline{59} \\
44\,8 \\
\underline{41\,3} \\
3\,5\,4 \\
\underline{3\,5\,4} \\
0
\end{array}
$$

(Quotient or answer)
(Dividend)

 103.84 ÷ 5.9 = 17.6

PRACTICAL PROBLEMS

Divide the following quantities. Round each answer to the number of decimal places given.

1. 0.96 ÷ 4 =   (two places) _____

2. 99.19 ÷ 0.7 =   (one place) _____

3. 8.75 ÷ 1.25 =   (one place) _____

4. 356.76 ÷ 62 =   (three places) _____

5. 0.0057 ÷ 19 =   (four places) _____

6. 0.0988 ÷ 0.012 =   (three places) _____

7. 73.275 ÷ 2.5 =   (two places) _____

8. 6.111 ÷ 0.97 =   (one place) _____

9. The dividers in one box weigh 16.32 lb. Each set of dividers weighs 0.68 lb. How many dividers are in the box?   _____

10. One thousand 45° triangles weigh 498.6 pounds. What is the weight of each triangle?

11. The diameter of a reamed hole is 70.36 mm. What is the radius setting, in millimeters, used to draw the hole?

12. The pitch of a screw thread is found by dividing 1.00 inch by the number of threads per inch. A 1 ¼ -13 ACME screw thread has 13 threads per inch. What is its pitch to the nearest thousandth?

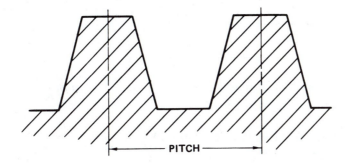

PITCH

13. A ¼" diameter screw has 28 threads per inch. Find, to the nearest thousandth inch, the pitch of the screw. (Pitch = 1 ÷ threads per inch).

14. All sections on this development are equal in length. Find, in millimeters, dimension **A**.

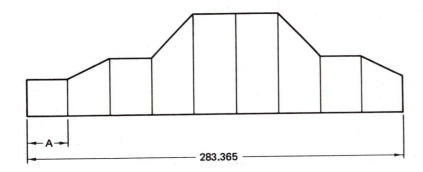

A

283.365

15. Four drafters work on a special job after regular working hours. They receive $726 to be shared equally. How much does each one receive?

16. All sections on the block are equal in length. Determine length **X** to the nearest thousandth of an inch.

_____

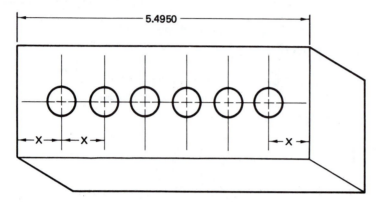

17. Seventeen drafters receive a total annual raise of $15,933.25. What equal amount does each one receive?

_____

18. The perimeter of this regular hexagon is 250.029 mm. Find, in millimeters, the length of each equal side.

_____

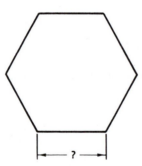

19. An object line on a detail drawing is 27.987 inches long. It is divided into 19 equal parts. Determine the size of each division.

_____

20. A drawing board is 0.750" thick. A stack of boards is 27.750" high. How many boards are in the stack?

_____

21. A CAD operator evenly assigns 135 levels to 9 different drawing types for a large building. Find the number of levels assigned to each type of drawing.

_____

22. A CAD drafter assigns 234 levels to 13 different types of drawings. Find the average number of levels assigned to each type of drawing.

_____

23. A civil drafter receives $826.50 for a 38-hour week. Find the drafter's rate of pay per hour.   _____

24. The CAD drawing shown is reduced by a factor of 4. Determine the new height and width. The grid spacing is currently 1.0".

Height _____

Width _____

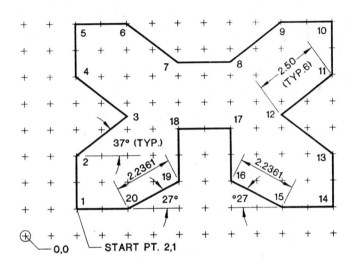

Reprinted with permission from Stellman, Krishnan, and Rhea,
Harnessing AutoCAD, copyright 1993 by Delmar Publishers.

25. Using the CAD drawing given, find the new size of the holes that must be reduced by a factor of 3 in order to work with another part.   _____

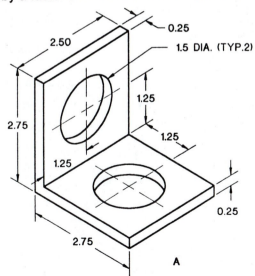

Reprinted with permission from Stellman, Krishnan, and Rhea,
Harnessing AutoCAD, copyright 1993 by Delmar Publishers.

26. The pressure safety valve is inserted in a CAD drawing with a reduction factor of 6. If the grid spacing currently is .75, find the new height and width of the valve.

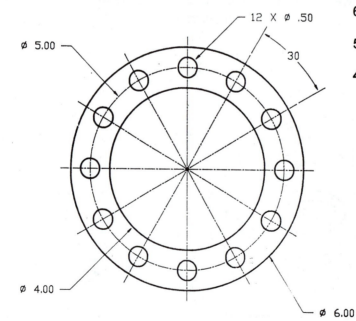

Reprinted with permission from Stellman, Krishnan, and Rhea,
Harnessing AutoCAD, copyright 1993 by Delmar Publishers.

Height _____

Width _____

27. A box of twelve 5 ¼" high density diskettes costs $22.44. What is the cost of one disk?

_____

28. The CAD drawing of the spacer ring can be modified to fit with another part if the large diameters are reduced by a factor of 1.5 and the smaller holes are reduced using a factor of 4. Determine the reduced sizes as indicated.

6.00 Diameter _____

5.00 Diameter _____

4.00 Diameter _____

.50 Diameter _____

29. A CAD drafter records the following number of hours devoted to architectural projects: 17.20 hrs., 15.75 hrs., 19.5 hrs., 16.4 hrs., 18.25 hrs., 11.0 hrs., and 14.30 hrs. What is the average number of hours spent on these projects?

          _____

30. Using the CAD drawing provided, determine the new values for dimensions **A–E** once the block has been scaled down by a factor of 3. Use three-place decimals with your answers.

A _____

B _____

C _____

D _____

E _____

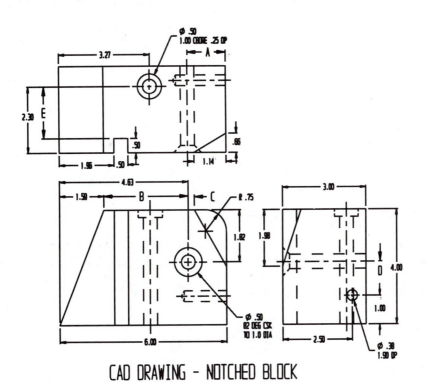

CAD DRAWING - NOTCHED BLOCK

Unit 15 DECIMAL AND COMMON FRACTION EQUIVALENTS

BASIC PRINCIPLES OF DECIMAL AND COMMON FRACTION EQUIVALENTS

Converting common fractions to decimal fractions

A common fraction is an algebraic expression for division. To change a fraction into its decimal equivalent, divide the numerator by the denominator.

Example: Change $\frac{5}{8}$ to a decimal.

$$
\begin{array}{r}
.625 \\
8\overline{\smash{)}5.000} \\
\underline{4\,8} \\
20 \\
\underline{16} \\
40 \\
\underline{40} \\
0
\end{array}
$$

Converting decimal fractions into common fractions

A decimal fraction is basically a common fraction having a denominator with a specific power of 10. The number of places to the right of the decimal point indicates the power of the denominator.

Examples: Express the following decimals as fractions.

$$.27 = \tfrac{27}{100}; \quad .063 = \tfrac{63}{1000}; \quad 3.9 = 3\tfrac{9}{10}$$

PRACTICAL PROBLEMS

Round all decimal fraction answers to the nearest thousandth (three places). Express the following common fractions as decimal fractions.

1. $\frac{2}{7}$ = _____

2. $\frac{15}{16}$ = _____

3. $\frac{13}{24}$ = _____

4. $\frac{19}{32}$ = _____

Express the following decimal fractions as common fractions.

5. 0.1875 = _____

6. 0.71875 = _____

7. 0.8125 = _____

8. 0.015625 = _____

9. The width of an erasing shield is $2\frac{7}{16}$". What is this width in decimal form? _____

10. Express, in inches, the dimensions on this locating finger as decimal fractions.

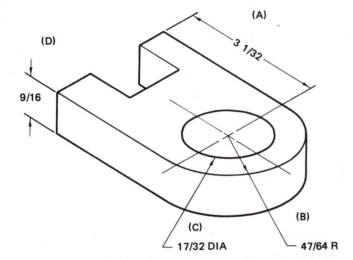

A _____

B _____

C _____

D _____

11. A civil drafter's dust brush weighs 1.25 lb. Find, in pounds, the weight of 23 brushes. Express the answer as a mixed number with a common fraction. _____

12. A machinist uses a lathe to turn down a $\frac{27}{64}$" diameter shaft from a $\frac{9}{16}$" diameter piece of stock. Find, in inches, the difference in diameters. Express the answer as a decimal. _____

13. Find, in inches, the perimeter of this die insert scale. Express the answer as a mixed number with a common fraction. _____

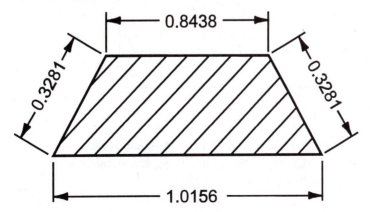

14. An eraser is 0.475 inches thick. What is the thickness written as a fraction?

_____

15. Find, in inches, the perimeter of this regular pentagon. Express the answer as a mixed number with a common fraction.

_____

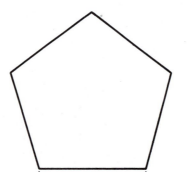

—0.875—

16. A line is divided into three segments of 3 $\frac{49}{64}$ ", 2 $\frac{13}{32}$ ", and 1 $\frac{19}{64}$ ". What is the total length of the line in inches? Express the answer as a decimal.

_____

17. Find, in inches, dimensions **A** and **B**. Express the answers using common fractions.

A _____

B _____

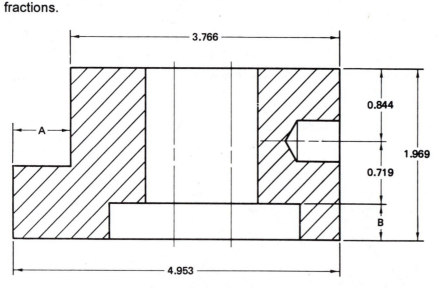

— 3.766 —

0.844

1.969

0.719

B

— 4.953 —

18. A visible object line on a drawing is approximately 0.034" thick. A hidden object line is approximately 0.018" thick. Find, in inches, the difference in the two thicknesses. Express the answer as a fraction.

_____

19. A 0.5312" diameter hole is drilled in the block shown before a tapping operation. The block is tapped with a ⅝ " diameter tap drill. Find, in inches, the difference in the two diameters. Express the answer as a common fraction.

_____

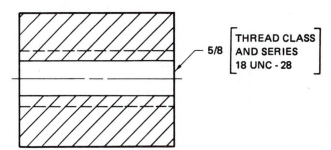

20. Find, in inches, dimension **A** on this stepped shaft. Express the answer as a common fraction.

_____

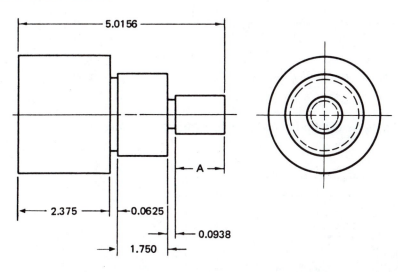

21. Find the tolerances for the holes contained on the shim presented. Express **A** as a decimal fraction and **B** as a common fraction.

A _____

B _____

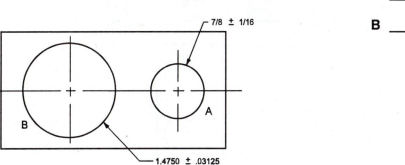

22.   Find dimensions **A, B, C,** and **D** using this illustration.

A _____

B _____

C _____

D _____

23.   Determine the difference decimally between a 5 ¼" and a 3 ½" floppy diskette.

_____

24.   Using the CAD drawing below, find dimension **A**. State your answer as a common fraction.

_____

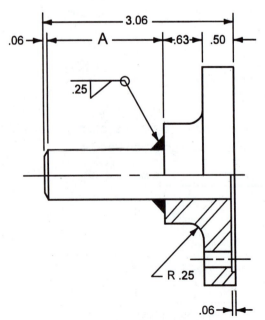

Reprinted with permission from
Resetarits and Bertolini,
Using CADKEY Light, copyright
1992 by Delmar Publishers.

25. The two-inch circles shown in the CAD drawing will be reduced by a factor of 7. Determine the new value and express your answer to the nearest standard common fraction. _____

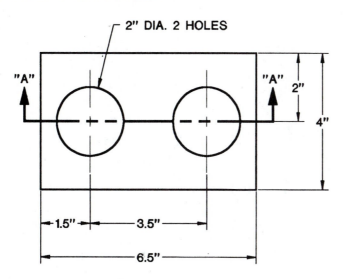

2" DIA. 2 HOLES

"A"   "A"   2"

4"

1.5"   3.5"

6.5"

26. Find dimensions **A** and **F** using the drawing shown. Dimension **B** is ¾ ", **C** is 1⅝ ", **D** is 2²⁹⁄₃₂ ", and **E** is 3⁵⁄₆₄ ". Express your answers as three-place decimal fractions.

A _____

F _____

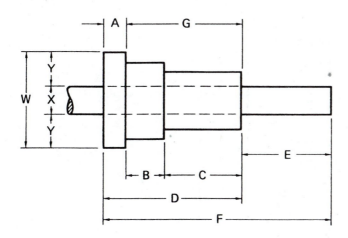

27. Using the CAD drawing given, find how far from the center of the large circle the 4 small holes will be if they moved out halfway between the two circular center lines (bolt circles). State your answer as a common fraction.

_____

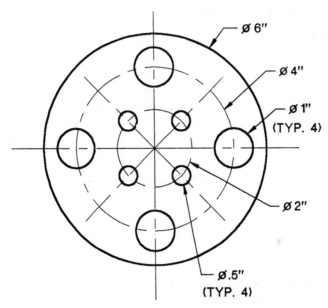

Ø 6"

Ø 4"

Ø 1"
(TYP. 4)

Ø 2"

Ø .5"
(TYP. 4)

28. Find dimensions **A** and **B** using the view given. Express your answers as decimal fractions.

A _____

B _____

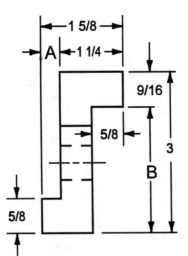

1 5/8

A   1 1/4

9/16

5/8

3

B

5/8

29. Determine the overall length and height of the symmetrical CAD drawing. State your answers as common fractions.

Length _____

Height _____

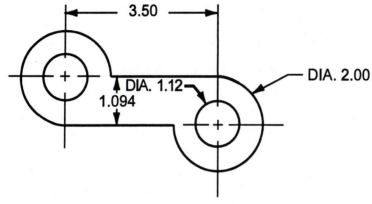

Reprinted with permission from Stellman, Krishnan, and Rhea,
Harnessing AutoCAD, copyright 1993 by Delmar Publishers.

30. Using the CAD drawing presented, find dimension **A** as a common fraction.

_____

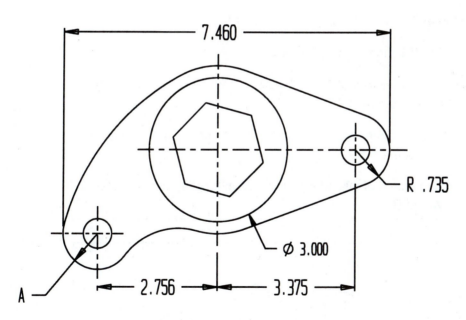

CAD DRAWING - ROCKER ARM

Unit 16 COMBINED OPERATIONS WITH DECIMAL FRACTIONS

BASIC PRINCIPLES OF COMBINED OPERATIONS

This unit contains practical problems involving combined operations of addition, subtraction, multiplication, and division with decimal fractions.

PRACTICAL PROBLEMS

Calculate each answer to the nearest thousandth.

1. Subtract 0.430 from 1.119 then multiply the difference by 0.256. _____

2. Multiply 0.98 by 0.673 then add 13.347 to the product. _____

3. Add 12.67 to 32.468 then divide the sum by 3.7. _____

4. Divide 114.64 by 6.3 then subtract the quotient from 42.86. _____

5. Add 17.63 to 39.257 then multiply the sum by 4.8. _____

6. Multiply 13.5 by 7.2 then divide the product by 5.2. _____

7. Two holes have diameters of 2.7625" and 1.3156". Find, in inches, the difference between the radii. _____

8. Three holes have diameters of 2.187", 0.965", and 6.373". Find, in inches, the total of the radii. _____

9. All sections of the arch of this bridge are equal in length. Find, in meters, dimensions **A** and **B**.

A _____

B _____

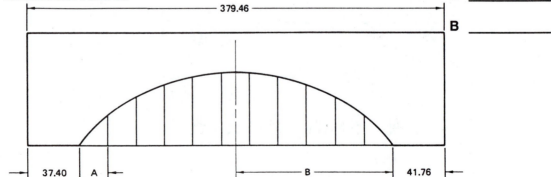

10. A block is 7.188" thick. Three rough cuts of 0.473" each and one finish cut of 0.009" are machined from the block. What is the remaining thickness of the block in inches? _____

11. The *circumference* of a circle equals the diameter times 3.1416. Find, in centimeters, the difference between the circumferences of the large and small circles. _____

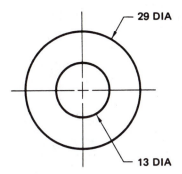

29 DIA

13 DIA

12. Artgum erasers are produced at a rate of 2,880 erasers per hour. How many erasers are made in 3 minutes and 45 seconds? _____

13. Find, in inches, dimensions **A** and **B** on this template. (**TYP** means that the same dimension is found at the other end.)

A _____

B _____

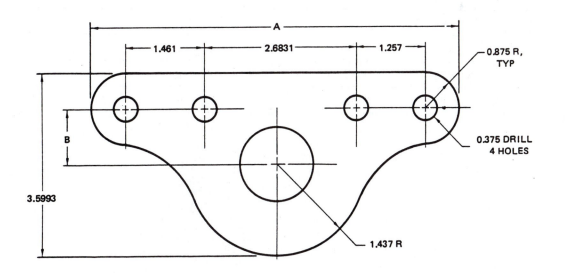

A

1.461   2.6831   1.257

0.875 R, TYP

0.375 DRILL
4 HOLES

B

3.5993

1.437 R

14. a. Find, in decimeters, the *smallest* allowable size for the hole in this drawing.
    _____

    b. Find, in decimeters, the *largest* allowable size for the hole in this drawing.
    _____

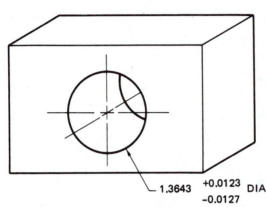

1.3643   +0.0123   DIA
          −0.0127

15. A supervisor makes a chart on the cost of drafting pencils that were purchased. •

    a. How much money was spent on the pencils?   _____

    b. How many pencils were purchased?   _____

TYPE OF PENCIL	COST PER DOZEN	AMOUNT BOUGHT	AMOUNT SPENT
F pencils	$2.68	3 dozen	
H pencils	$2.78	1.5 dozen	
2H pencils	$3.02	2 dozen	
6H pencils	$3.24	2.25 dozen	
		TOTAL COST:	

16. Find, in millimeters, dimensions **A** and **B** on this magnet.

A _____

B _____

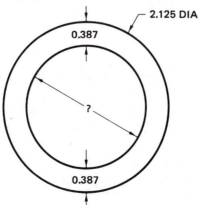

17. Find, in inches, the overall length of this match plate and dimension **A**.

Overall
length _____

A _____

0.250 DIA., 15 HOLES,
EQUALLY SPACED AT 0.875

0.39

◄A►

◄0.63►

?

18. The outside diameter *(OD)* of a section of pipe is 2.125". The wall
thickness is 0.387". Find, in inches, the inside diameter *(ID)* of the pipe.   _____

2.125 DIA

0.387

?

0.387

19. Three principal views of an object are centered on a 12 × 18 sheet of vellum. Find, in inches, dimensions **A** and **B**.

A _____

B _____

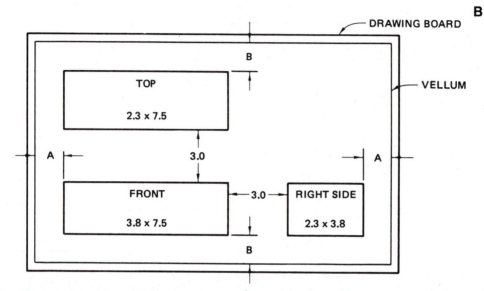

20. Two drafting departments work on a large job. Department *A* works 56.5 hours, 87.8 hours, 106.7 hours, 37.4 hours, and 77.3 hours. Department *B* works 97.9 hours, 62.7 hours, 48.3 hours, and 102.75 hours. How many more hours does Department *A* work than Department *B*?

_____

21. A small CAD department purchased the following items: a stand-alone computer $5,225.50, an optical mouse $97.36, a plotter $1,119.63, a dot-matrix printer $647.87, and CAD software $432.44. (a) Find the total cost of this CAD station. (b) What would be the total excluding the printer? (c) What would it cost to equip 5 CAD drafters with a complete CAD station?

a. _____

b. _____

c. _____

22. The circumference of a circle equals the diameter times 3.1416. Find the difference between the circumference of a 3.75 diameter pulley and the circumference of a 9.25 diameter wheel. Express the answer to the nearest hundredth.

_____

23. A mechanical drafter finds a dimension of 3 5/7 inches on a drawing. What standard common fraction is closest to this dimension?

_____

24. Find dimensions **A** and **B** using the CAD drawing provided. State the answers as decimal fractions.

A _____

B _____

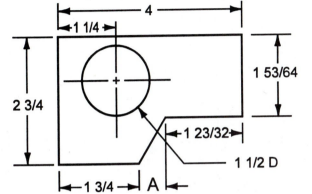
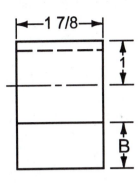

Reprinted with permission from Resetarits and Bertolini,
Using CADKEY Light, copyright 1992 by Delmar Publishers.

25. Using the CAD drawing and the dimensions provided, find the overall height and width.

Height _____

Width _____

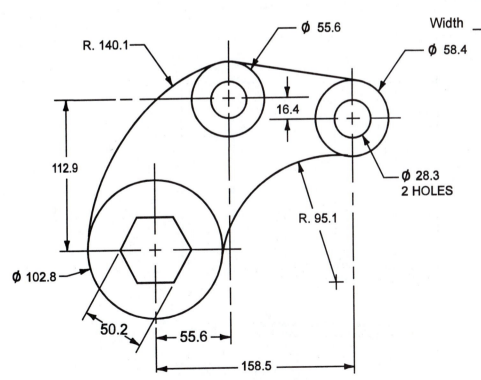

Reprinted with permission from Resetarits and Bertolini,
Using CADKEY Light, copyright 1992 by Delmar Publishers.

26. The height of the CAD drawing presented will be reduced by a factor of 5, the width by a factor of 3, when the object is inserted into another CAD drawing. What will be the new height and width dimensions?   Height _____

Width _____

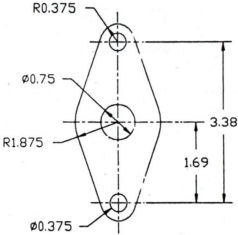

R0.375

Ø0.75

R1.875

Ø0.375

3.38

1.69

Reprinted with permission from Resetarits and Bertolini,
Using CADKEY Light, copyright 1992 by Delmar Publishers.

27. Determine the area of the CAD drawing that is shown using a .375" grid.   _____

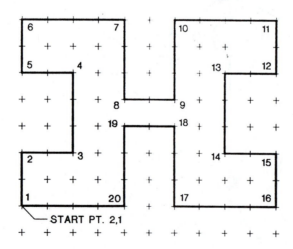

START PT. 2,1

Reprinted with permission from Stellman, Krishnan, and Rhea,
Harnessing AutoCAD, copyright 1993 by Delmar Publishers.

28. Using the CAD drawing illustrated, find the new overall height and width
    if the drawing is inserted and enlarged using a scale factor of 1.33.   Height _____

    Width _____

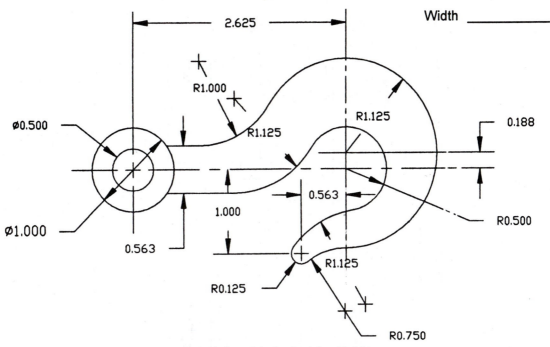

Reprinted with permission from Resetarits and Bertolini,
Using CADKEY Light, copyright 1992 by Delmar Publishers.

29. The CAD drawing below is set up using a .875" grid. Find its perimeter.   _____

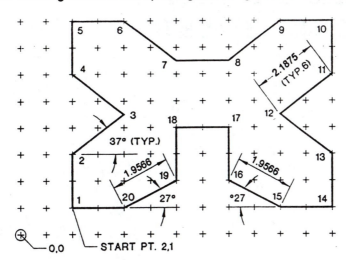

Reprinted with permission from Stellman, Krishnan, and Rhea,
Harnessing AutoCAD, copyright 1993 by Delmar Publishers.

30. Find dimensions **A**, **B**, and **C** using the CAD drawing shown. Determine
    the three overall dimensions if the object is scaled down by a factor of 4.

    A _____

    B _____

    C _____

    Height _____

    Width _____

    Depth _____

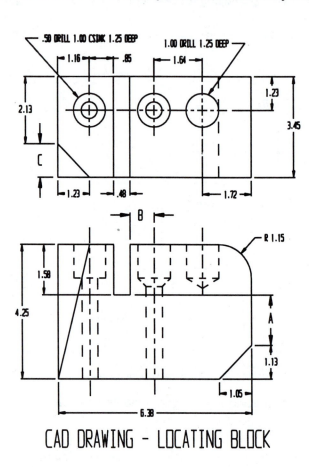

CAD DRAWING - LOCATING BLOCK

Percent, Averages, and Estimating

Unit 17 PERCENT AND PERCENTAGE

BASIC PRINCIPLES OF PERCENT AND PERCENTAGE

Percent means hundredths or a way to say "How many parts per hundred." For example, if a CAD drafter worked 25 hours of the hundred hours devoted to a drafting project, 25% of the work was accomplished by the CAD drafter. Percent may be expressed as a decimal by first changing the number to a fraction and then to a decimal, then move the decimal point two places to the left.

Example: $43\% = \frac{43}{100} = 0.43$

▦ 43 ÷ 100 = 0.43

To change a percent to a fraction, place the percent over 100 and reduce the fraction to its lowest terms.

Example: $40\% = \frac{40}{100} = \frac{4}{10} = \frac{2}{5}$

In solving percentage problems, change the percent to a decimal, then treat "of " as "times."

A basic formula to calculate percentage is:

Part = Percent × Whole
$$P = \% \times W$$

Examples: Find 25% of 24 inches

$.25 \times 24 = 6$ inches

What percent of 60 feet is 15 feet?

$\% = \frac{P}{W}$ $\% = \frac{15}{60} = 0.25 = 25\%$

48 is 75% of what number?

$W = \frac{P}{\%}$ $W = \frac{48}{.75}$ $W = 64$

PRACTICAL PROBLEMS

1. 7% of 48 _____

2. 27% of 86 _____

3. 15% of 75 _____

4. 35% of 197 _____

5. 21% of 1192 _____

6. 87 $\frac{1}{4}$ % of 724 _____

7. A checker in a structural drafting department checked 150 drawings in one month. There were more than 12 mistakes on 22% of the drawings. How many drawings had more than 12 mistakes? _____

8. Shade in 75% of the illustrated figure. _____

9. A CAD drafter predicts that a specific job should take 150 hours. The number of hours to complete the job is 7% less than the predicted time. Find the number of hours the CAD drafter takes to complete this job. _____

10. A machined block 3.345 in. thick must be reduced by 18%. What is the new thickness after the reduction? _____

11. An architectural drafting department supervisor orders 136 reams of assorted sizes of paper. When the order is delivered, 12 $\frac{1}{2}$ % of the total order is back ordered and is not delivered. Determine how many reams of paper are not delivered. _____

12. Using the illustrated figure, how many blocks would be shaded to represent 37.5%? _____

13. A CAD drafter invests 35% of $10,000 savings in a piece of property. Determine the amount of money invested in this property. _____

14. A designer must redesign the plate shown so that dimension **A** is 3% smaller and dimension **B** is 9% larger. What are the new dimensions of **A** and **B**? Round answer to the nearest thousandth inch.

A _____

B _____

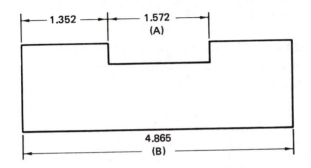

15. Assume that a freehand sketch can be made in $\frac{1}{10}$ of the time it takes to make an instrument drawing. If an instrument drawing takes $2\frac{1}{2}$ hours, how many minutes should a freehand sketch take? _____

16. Determine how many sections of the circle should be shaded to represent 75% of the circle. _____

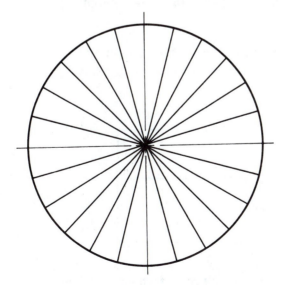

17. A large drafting department had 368 drafters. Over a two-year period 6 ¼ % of the drafters retired. How many of the original drafters are working in the department?

_____

18. A drafting job was completed with standard drafting equipment in 60 hours. A drafting machine with a protractor head is said to decrease the drafter's time by 30%. Based on this figure, how many hours would have been saved if the drafter used a drafting machine?

_____

19. Due to an incorrect dimension on a detail drawing, a drafter must reduce the hole diameter in this stop block by 13%. What is the new hole size? Round answer to the nearest thousandth inch.

_____

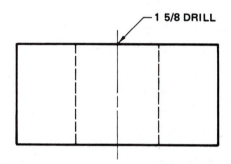

20. Of the 150 drafters employed by a company, 18% have part-time jobs. How many have part-time jobs?

_____

21. The overall length of the T guide illustrated must be enlarged by 12 ½ %. What is the new length? Round answer to the nearest thousandth inch.

_____

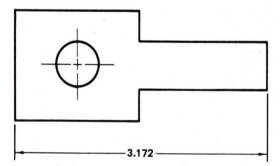

22. Find the average thickness of several pieces of sheet metal which measure: 0.313 in., 0.267 in., 0.417 in., 0.135 in., 0.547 in., 0.226 in., and 0.343 in. Round the answer to the nearest thousandth of an inch.

_____

23. A large engineering firm employs 76 drafters. There are 12 structural drafters, 19 mechanical drafters, 14 civil drafters, 25 architectural drafters, and CAD drafters. How many CAD drafters work for this company? Find the approximate percentage for each classification of drafter. Round answers off to whole numbers.

CAD Drafters	_____	% Structural	_____
% Mechanical	_____	% Civil	_____
% Architectural	_____	% CAD	_____

24. A CAD drafter assigns 200 layers of the 256 available in the following manner: 75 for plan views, 50 for elevations, 40 for details, 25 for construction, and 10 for miscellaneous specifications. What percentage of the layers were used? Find the percentage for each layer assigned from the total of layers used.

% of Layers Used	_____	% Plan Views	_____
% Elevations	_____	% Details	_____
% Construction	_____	% Miscellaneous	_____

25. A CAD drafter must enlarge the CAD drawing presented so that it can fit on top of another part. The height will be enlarged by 20% and the width by 15%. Determine the new height and width dimensions.

Height _____

Width _____

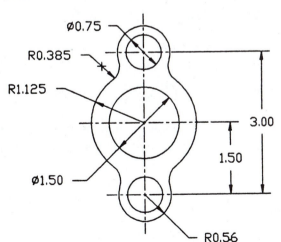

∅0.75

R0.385

R1.125

3.00

1.50

∅1.50

R0.56

Unit 18 INTEREST AND DISCOUNTS

BASIC PRINCIPLES OF INTEREST AND DISCOUNTS

Interest and discounts are ways of using percentages. First, let's look at the concept of interest. When money is borrowed, the borrowed amount is known as the principal. Interest is a payment for the use of the borrowed money and the rate of interest is expressed as a percent. Interest is usually calculated on the basis of one year. When the term of the loan has expired, the money repaid equals the principal plus the interest.

Formulas

1. Simple interest = principal × rate × time (in years) (I = PRT)
2. Amount = Principal + Interest

Example:

A company borrows $15,000.00 at a rate of 12% over a two-year period. How much money is repaid after two years?

I = PRT   I = $15,000 × 12% × 2
            I = $15,000 × 0.12 × 2
            I = $3,600

         $15,000 + $3,600 = $18,600 = amount repaid

         🖩   15,000 Ⓧ 12 ⑨ Ⓧ 2 ⊜ 3,600 ⊕ 15,000 ⊜ 18,600

Discount rate is a percent by which a price is marked down. The list price multiplied by the discount rate equals the discount. (L × R = D).

The list price minus the discount equals the net price (sales price) (L − D = N).

Example: A set of drafting instruments lists for $38.00. It has been marked down 10%. An additional 3% is given if paid for in cash. Find the cost if a buyer pays in cash.

         $38.00 × 0.10 = $3.80
         $38.00 − $3.80= $34.20
         $34.20 × 0.03 = $1.03
         $34.20 − $1.03= $33.17 = cost for cash buyer

PRACTICAL PROBLEMS

1. The list price of a computer is $4,800.00 with a discount of 10% off this price. Find the cost of the computer.   _____

2. The list price of a computer program is $178.00 with a 15% discount off list price and a cash discount of 2%. Find the net cost.   _____

3. A civil engineer purchases an item through a wholesale outlet paying 85% of the retail price. What percent of the retail price is saved?   _____

4. Determine the cost of four lengths of steel as illustrated. The steel weighs 2.5 pounds per foot and costs $1.48 per pound. (The dimension is in inches.) Find the cost if a cash discount of 9% is applied.

   Cost _____

   Discount price _____

   |— 30 —|

5. A package of 250 sheets of drafting vellum costs $17.50. Find the cost of each sheet of vellum. What is the cost of 12 packages of vellum with a discount of 7%?

   Cost _____

   Discount price _____

6. The cost of a drafting instrument set is $24.65. If the buyer pays cash, a 5 $\frac{1}{2}$ % discount is given. Determine the net price, to the nearest cent, if the buyer pays cash for the instrument set.   _____

7. The list price for T squares is $40.44 per dozen. There is a trade discount of 19%. Find the cost, to the nearest cent, of each T square when a company buys a dozen.   _____

8. The total cost of a drafting subcontractor's job is $426.00. The materials cost 6% of the total amount, computer time cost 14% of the total amount, and the rest was charged to labor. What are the amounts of each of the three costs for the job?

   Materials _____

   Computer time _____

   Labor _____

9. If 1,440 wedge blocks can be stamped out in one hour, determine how many are made in 45 minutes and 15 seconds.

_____

10. A 23% reduction is made on $176.56. Find the price, to the nearest cent, with the discount.

_____

11. A manufacturer figures the following costs on a drafting project: $27.50 for materials, $425.00 for labor, and $135.00 for overhead. The profit is 37% of the total cost. What must be charged to the customer to get this profit? Round answer to the nearest cent.

_____

12. The special nozzle fitting sells for $42.72 per dozen less discounts of 10%, 6%, and 2%.

a. Determine the discount price, to the nearest cent, of a dozen fittings.

_____

b. Find the price of each fitting. Round the answer to the nearest cent.

_____

13. The total cost of a drafting package includes the cost of materials ($12.68), labor charge ($1,255.00), and the overhead costs ($375.00). Find, to the nearest whole percent, the percent of the total cost represented by each of the following.

Materials _____

Labor _____

Overhead _____

14. Calculate the cost of 3,600 pounds of metal at 38¢ per pound less discounts of 12%, 9%, and 3%.

_____

15. The cost of manufacturing 1,000 shims as illustrated is $863.00. Of this, 31% is spent for materials, 48% for labor, and the remainder for overhead.

a. Find the cost for materials.

18 GAUGE

_____

b. Find the cost for labor.

_____

c. Find the cost for overhead.

_____

16. The list price of a complete set of drafting instruments is $36.74. The set is sold with a 14% discount. What is the net cost of the instrument set? _____

17. A company purchased 24 set-up blocks like the one illustrated for use in the tool room. The set-up blocks sell for $85.60 for a set of four. At the time of purchase a 12% trade discount and a 4% cash discount were allowed. _____

    a. Find the cost of 24 set-up blocks before the discount. _____

    b. Find the cost of 24 set-up blocks when a company makes a cash purchase. _____

    c. Determine the cost to the company for each set-up block purchased. _____

18. The payroll for a small drafting department is $946. Find the percent of the payroll spent in the drafting department when the payroll for the company is $4,730. _____

19. The average yearly operating cost of a drafting department is $135.00 per person. Find the operating cost for the year if the drafting department has 57 workers. _____

20. The cost of manufacturing the step shaft illustrated is reduced 17% by eliminating the undercuts at the shoulders. The original cost is $9.37.

UNDERCUTS ELIMINATED
BOTH SIDES

    a. Find the savings per individual part. _____

    b. Find the new cost of producing the step shaft. _____

21. How much interest will $4,500.00 earn in one year if the interest rate is $6\frac{3}{4}$%?

_____

22. A loan is taken out for $54,000.00 at 12% per annum to enlarge the drafting department. If the loan is repaid in 18 months, how much interest is charged?

_____

23. A loan is secured for $7,600.00 at 8% per annum. The note is paid when it becomes due in 27 months. Find the amount of interest.

_____

24. The annual rate of interest on $1,500 is 17%. What is the interest after a period of 3 months? Find the interest after a period of 4 months.

_____

25. Find the amount paid on a loan of $8,600 at 15% after 3 years.

_____

26. A bank loans a drafter $4,500 on which $675.00 interest is paid annually. What is the rate of interest?

_____

27. A family pays $5,490 interest annually on a mortgage of $36,000. What is the rate of interest?

_____

28. Find the interest for 1 year, at $9\frac{1}{2}$% on $3,735.75.

_____

29. A bank note is issued for $3,725 at a rate of 1% for one month. How much interest is charged for 3 months?

_____

30. An engineering firm adds an 18% interest charge (1.5% per month) on money owed them after 30 days. How much interest is owed on a bill of $3,750 which is paid in 90 days?

_____

Unit 19 AVERAGES

BASIC PRINCIPLES OF AVERAGES

Averages are found by dividing the sum of the numbers in a set by the number of members in the set.

Example: Find the average of the following numbers: 46, 17, 21, and 16.

$$
\begin{array}{r}
2 \\
46 \\
17 \\
21 \\
+\ 16 \\
\hline
100
\end{array}
\qquad
\begin{array}{r}
25 \\
4\,)\overline{100} \\
\underline{100} \\
0
\end{array}
\qquad 25 = \text{average}
$$

46 ⊞ 17 ⊞ 21 ⊞ 16 ⊟ 100 ⊡ 4 ⊟ 25

PRACTICAL PROBLEMS

1. Three CAD drafters receive hourly wages of $12.35, $10.85, and $11.55 respectively. What is the average hourly pay?

   _____

2. A CAD drafter assigns the following number of levels to certain phases of engineering project: 9, 12, 26, 42, 37, 22, and 20. Find the average number of levels that were assigned to this project.

   _____

3. Over a five-day period, the following work-related miles are logged: 48, 33, 28, 40, and 37. Find the average number of miles traveled per day.

   _____

4. A mechanical drafter completes the following number of drawings over a period of four working weeks: 19, 17, 16, and 12. What is the weekly average?

   _____

5. A class of drafting students made the following scores on their first exam: 78, 98, 82, 66, 92, 83, 79, 87, and 82. What is the average score on this exam?

   _____

6. An inexperienced CAD drafter earns $12.50 per hour. During a five-day period, these hours are worked: 8, 5 ½, 9, 6 ¼, and 7 ¾ hours. Find the average daily earning.

   _____

141

7. Find the average for the following set of fractions: ⅔, ⁵⁄₁₆, ⁷⁄₁₂, and ¾.   _____

8. A detail drafter completes 84 drawings over a one-year period. Find the drafter's monthly average.   _____

9. An architect designs a building with 5 different types of doors. The doors cost $185.50, $343.75, $1,255.30, $737.45, and $279.65. Find the average cost per door.   _____

10. On a drafting test, the scores achieved are as follows: 2 students, 87%; 3 students, 76%; 4 students, 93%; 5 students, 81%; and 1 student, 67%. What is the average score on the test?   _____

11. A CAD drafter completes 11 drawings on Monday, 9 on Tuesday, 13 on Wednesday, and 17 on Thursday. How many must be completed on Friday to average 13 drawings per day?   _____

12. What is the average length of the dimensions shown on the template? All dimensions are in millimeters. Round answer to the nearest hundredth of a millimeter.   _____

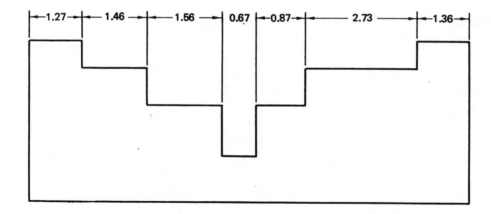

13. A civil drafter spent the following amounts for equipment during the year: $16.20, $35.78, $78.36, $42.83, and $56.52. What is the average of these payments? Round answer to the nearest cent.   _____

14. Over a period of seven years, a drafter misses the following working days per year: 16, 7, 11, 9, 7, 5, and 8. What is the average number of working days missed per year?   _____

15. What is the average weight of six bars of metal which weigh: $6\frac{1}{2}$ lb., $3\frac{1}{4}$ lb., $2\frac{7}{8}$ lb., 7 lb., $5\frac{3}{4}$ lb., and $12\frac{3}{8}$ lb.? State your answer in decimal form correct to the nearest hundredth of a pound.

_____

16. What is the average length, in inches, of the lines shown in the illustration? State your answer in decimal form correct to the nearest thousandth of an inch.

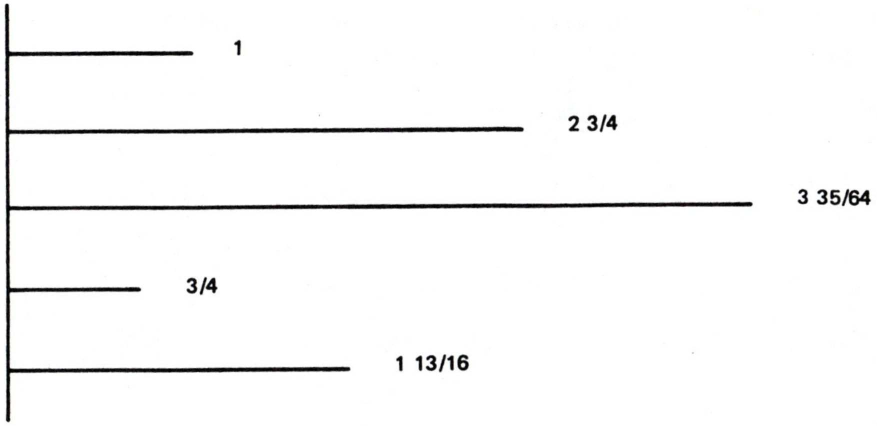

_____

Unit 20   ESTIMATING

BASIC PRINCIPLES OF ESTIMATING

Estimates are used for approximation purposes and are not intended to be exact. Estimating the amount of the time needed to complete a task or the cost of materials requires time and experience.

Example: An architect estimates that it will take 56 hours to complete a proposed design for a customer. The design was finished in 52 hours. The actual amount is within what percent of the estimate?

$$52 \div 56 = 93\% \qquad 100\% - 93\% = 7\%$$

Alternate solution: Find the error,
$$56 - 52 = 4 \text{ hours} \qquad 4 \div 56 = 7\%$$

The actual amount is within 7% of the estimated amount.

    56 ⊖ 52 ⊜ 4 ⊙ 56 ⊜ .07 = 7%

PRACTICAL PROBLEMS

1. An office supervisor estimates that $135.00 will be spent on supplies next month. When the bill for supplies arrives, the cost is only $123.00. Find the percent of accuracy of the estimate.   _____

2. An architectural drafter estimates that 180 windows will be needed in a building. It is found that only 165 windows were needed. Find the percent of accuracy of the estimate.   _____

3. A structural drafter estimates that 50 gusset plates will be needed to frame a building. It was found that only 37 were needed. The actual amount is within what percent of the estimate?   _____

4. A civil drafter estimates a parcel of land contains 11,640 square feet. When surveyed, the parcel of land contains 10,980 square feet. Find the percent of accuracy of the estimate.   _____

5. A mechanical drafter estimated it will take 128 hours to complete a project. Electrical schematics take 16 hrs.; an assembly drawing, 31 hrs.; casting drawings, 26 hrs.; and machined parts, 47 hrs. The actual amount is within what percent of the estimate?   _____

6. A CAD drafter estimates it will take 60 hrs. to complete a drafting project. It took 66 hrs. to finish the project. What was the percent of accuracy of the estimate?   _____

7. An architect estimates the square footage of a building that will be used for manufacturing will be 12,750 square feet. The owner actually uses 11,300 square feet of space for manufacturing purposes. The actual amount is within what percent of the estimate?   _____

8. A drafting department supervisor estimates that 500 hours of drafting will be spent during a specific month. In fact, 535 hours are logged for that month. Find the percent of accuracy of the estimate.   _____

9. A CAD drafter estimates that 35 symbols were used on a floor plan of a large building. When counted, 47 symbols were actually used. The actual amount is within what percent of the estimate?   _____

10. An architect estimates that it will take 120 bundles of roofing shingles to cover the roof of a building. When the job is completed, it is found that 113 bundles were used. What is the percent of accuracy of the estimate?   _____

Unit 21 TOLERANCES

BASIC PRINCIPLES OF TOLERANCES

Tolerancing of dimensions is necessary when the dimensions of manufactured parts must be held to a specific degree of accuracy. Interchangeable manufacturing allows parts to be made anywhere, but when assembled they will function as designed. Tolerances and allowances placed on products make this possible.

A tolerance is a permissible variance in the size of a product or the difference between the maximum and minimum limits of a dimension.

Examples: Express 1.500 ± .003 as a limit dimension and then find its tolerance.

$$1.500 \pm .003 \text{ as limits } = \quad \underline{1.503} \qquad \text{High} \quad 1.503$$
$$1.497 \qquad \text{Low} \; - \; \underline{1.497}$$
$$.003 \; = \text{ Tolerance}$$

$$1\tfrac{1}{4} \pm \tfrac{1}{16} \text{ as limits } = \quad 1\tfrac{5}{16} \qquad \text{High} \quad 1\tfrac{5}{16}$$
$$1\tfrac{3}{16} \qquad \text{Low} \; - \; 1\tfrac{3}{16}$$
$$\tfrac{2}{16} \; = \; \tfrac{1}{8} \; = \text{ Tolerance}$$

Tolerances can vary in two directions (bilaterally) from the basic size.

Examples: ± .006
.006 × 2 = .012   Tolerance = .012

+ .003   .003
− .002   +.002
               .005   Tolerance = .005

Tolerances that vary in only one direction from the basic size are called unilateral tolerances.

Examples: + .003
− .000   Tolerance = .003

+ .000
− .007   Tolerance = .007

An allowance is the intended difference between the size of mating parts such as shafts, pulleys, or wheels. The allowance is the difference between the smallest hole size and the largest shaft size. A clearance fit provides clearance between mating parts and has a positive allowance. An interference fit, on the other hand, results in an interference between mating parts and the allowance is always negative.

Examples: Find the tolerances and the allowance for the mating parts expressed as limit dimensions.

Hole in Pulley	Smaller	.500	.502
	Larger	.502	− .500
			.002 = Tolerance

Diameter of Shaft	Smaller	.498	.498
	Larger	.495	− .495
			.003 = Tolerance

Allowance =   .500   (smallest hole size)
              − .498   (largest shaft size)
                .002   = Allowance (clearance fit)

Hole in Wheel   .750
                   .753   Tolerance = .003

Diameter of Shaft .758
                   .756   Tolerance = .002

Allowance =   .758   (largest shaft)
              − .750   (smallest hole)
                .008   (Interference)

Allowance = −.008 (minus sign shown)

PRACTICAL PROBLEMS

1. A dimension is given as $.856/_{.827}$. Find the tolerance on this dimension.   _____

2. Find the tolerance on the dimension stated as 1.125 ± .007.   _____

3. Determine the limits and tolerance on the dimensional value 4.375 ± $.015/_{.009}$.

Upper limit _____

Lower limit _____

Tolerance _____

4. What is the tolerance on the dimension provided?

    $3\frac{1}{2} + \frac{1}{16}$
    $\quad - \quad 0$

    _____

5. Find the lower limit on a dimension having a bilateral tolerance of ± .023 and an upper limit of 2.612.

    _____

6. Determine the limits for a dimension given as $3\,\frac{5}{16} \pm \frac{1}{64}$.   Upper limit _____

    Lower limit _____

7. Determine the upper limit on a dimension having a unilateral tolerance of +.017 and a lower limit of 5.125.

    −.000

    _____

8. Determine the limits for a dimension given as $1\,\frac{1}{2} + \frac{1}{32}$
    $\quad\quad\quad\quad\quad - \frac{1}{64}$ Upper limit _____

    Lower limit _____

9. Find the tolerances for a hole dimension of .6250 and a shaft dimension of .6240.
    .6233   .6255

    Hole _____

    Shaft _____

10. What are the tolerances and allowance for a 2.2500 diameter hole and a shaft diameter of 2.2485?   2.2512
    2.2472

    _____

    _____

11. Find the allowance and tolerances for the parts shown. What type of fit is this?   Hole tolerance _____

    Shaft tolerance _____

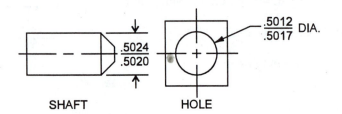

.5012 / .5017 DIA.

.5024 / .5020

SHAFT   HOLE

    Allowance _____

    Type of fit _____

12. Determine the limits and allowance for the examples shown. What type
    of fit results when these two parts are assembled?

Upper limit _____

Lower limit _____

Allowance _____

Type of fit _____

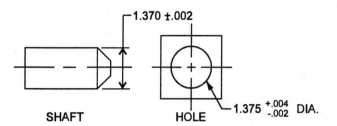

SHAFT   HOLE   1.375 $^{+.004}_{-.002}$ DIA.

1.370 ±.002

13. Determine the allowance for a hole dimension of 1.625 +.002 and a
    shaft dimension of 1.635 +.003.   −.003
                          −.002

_____

14. Find the limits for the shaft if a clearance fit is to result. The hole
    dimension is 1.500 ± .003 and the allowance is .007. The tolerance on
    the shaft is .005.

_____

15. Find the limits for the hole if an interference fit is to result. The shaft
    dimension is 1.375 +.009 and the allowance is .027. The tolerance on
    the hole is .009.   −.003

_____

16. Using the CAD drawing illustrated, determine the tolerances applied to
    dimensions **A** and **B**. Find the internal diameter size for the part that will
    mate with dimension **A** (at its largest size), providing an allowance of
    .0037".

Tolerance—**A** _____

Tolerance—**B** _____

Diameter
mating
part _____

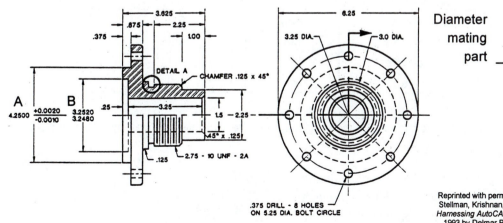

.375 DRILL - 8 HOLES
ON 5.25 DIA. BOLT CIRCLE

Measurement

 ## Unit 22 LINEAR MEASURE

BASIC PRINCIPLES OF LINEAR MEASURE

Linear measure refers to "straight line measurement".

A drafter must often express given units as larger or smaller measurements when laying out and dimensioning drawings. When the drawing is being made from the actual object, micrometers and vernier calipers are often used. The use of all drafting equipment requires accuracy and precision with both English and metric systems of measurement.

Lengths of English and metric measure are presented below.

COMMON ENGLISH LINEAR UNITS		
12 inches	=	1 foot
3 feet	=	1 yard
16 $\frac{1}{2}$ feet	=	1 rod
5,280 feet	=	1 mile

COMMON METRIC LINEAR UNITS	SYMBOL	VALUE IN METERS
1 millimeter	mm	0.001
1 centimeter	cm	0.01
1 decimeter	dm	0.1
1 meter	m	1.0

When converting feet and inches to inches, multiply the number of feet by 12, then add the number of inches to the product.

Example: Express 3 feet 7 $\frac{3}{4}$ inches as inches. (3'-7 $\frac{3}{4}$ ")
  3 × 12 = 36   3 feet = 36 inches
  36 inches plus 7 $\frac{3}{4}$ inches = 43 $\frac{3}{4}$ inches

To express inches as feet and inches, divide the number of inches by 12 to find the number of feet. The remainder equals the number of inches.

Example: Express 145 $\frac{1}{2}$ inches as feet and inches.

$$
\begin{array}{r}
12 \text{ (feet)} \\
12 \overline{)\ 145\ \ \frac{1}{2}} \\
\underline{12} \\
25 \\
\underline{24} \\
1 \ \ \frac{1}{2} \text{ Remainder (inches)}
\end{array}
$$

Answer: 12'-1 $\frac{1}{2}$ "

To express feet and inches as a decimal fraction in inches to the nearest hundredth, find the total number of inches and convert to a decimal. Fractions of an inch are converted to decimal form as well.

Example: Express 2 feet 7 $\frac{1}{2}$ inches to a decimal fraction in inches to the nearest hundredth.

2 × 12 = 24 inches

24 + 7 = 31 inches

$\frac{1}{2}$ = .50

Answer: 31.50 inches

PRACTICAL PROBLEMS

1. Express in inches.

    a.   2 feet = _____

    b.   3 feet 9 inches = _____

    c.   4 feet 7 $\frac{3}{4}$ inches = _____

    d.   5 feet 3 $\frac{1}{4}$ inches = _____

2. Express in feet and inches.

    a.   7.25 yards = _____

    b.   12 $\frac{1}{2}$ yards = _____

    c.   8.75 yards = _____

    d.   13 $\frac{1}{3}$ yards = _____

3. Express each of the following as decimal fractions in inches to the nearest hundredth.

    a.   3 feet 6 inches = _____

    b.   5 feet 11 $\frac{1}{4}$ inches = _____

    c.   7 feet 4 $\frac{7}{16}$ inches = _____

    d.   9 feet 8 $\frac{3}{8}$ inches = _____

4. Determine the readings indicated on this English scale.

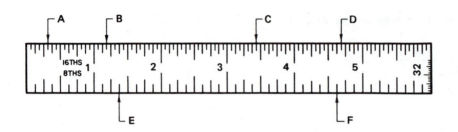

A _____

B _____

C _____

D _____

E _____

F _____

5. Determine the readings indicated on this English scale.

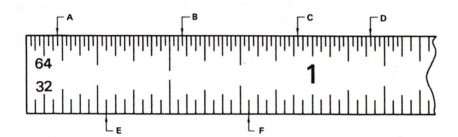

A _____

B _____

C _____

D _____

E _____

F _____

6. Determine the readings indicated on this English scale.

A _____

B _____

C _____

D _____

E _____

F _____

7. Determine the readings on this metric scale.

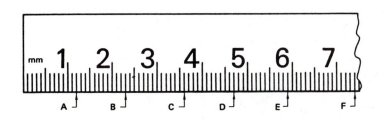

A _____

B _____

C _____

D _____

E _____

F _____

8. Read and record the settings on the 0.001 inch micrometer scales shown here.

Examples:

Reprinted with permission from Smith, *Vocational-Technical Mathematics*, 3e, copyright 1996 by Delmar Publishers

Answer = .857"   Answer = .263"

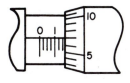

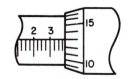

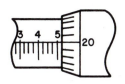

a. _____

b. _____

c. _____

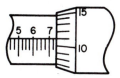

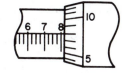

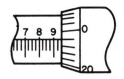

d. _____

e. _____

f. _____

9. Read and record the vernier caliper measurements on the scales shown here.

Example:

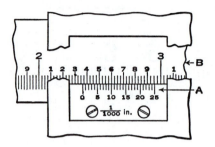

Reprinted with permission from Smith, *Vocational-Technical Mathematics*, 3e, copyright 1996 by Delmar Publishers

Answer = 2.359"

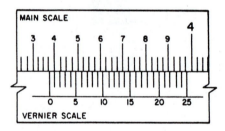

a. _____

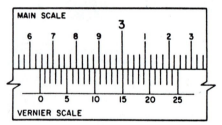

b. _____

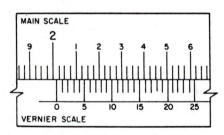

c. _____

10. Determine the readings on the ¼" architect's scale.

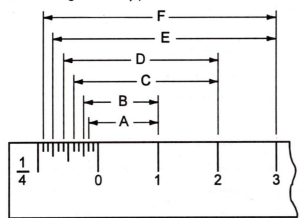

A _____

B _____

C _____

D _____

E _____

F _____

11. Determine the readings on the ⅜" architect's scale.

A _____

B _____

C _____

D _____

12. Determine the readings on the ¾" architect's scale.

A _____

B _____

C _____

D _____

E _____

F _____

13. Measure each line using the scale indicated and record its length in the space provided. Be as accurate as the specific scale will allow.

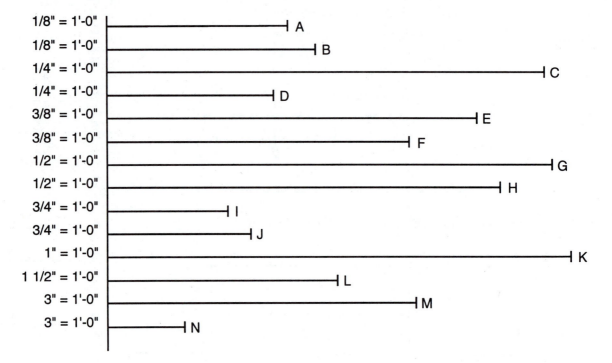

a. _____   b. _____   c. _____   d. _____

e. _____   f. _____   g. _____   h. _____..

i. _____   j. _____   k. _____   l. _____

m. _____   n. _____

14. Using an engineer's scale, measure each line to the nearest tenth.
     Example: 12.3 feet.

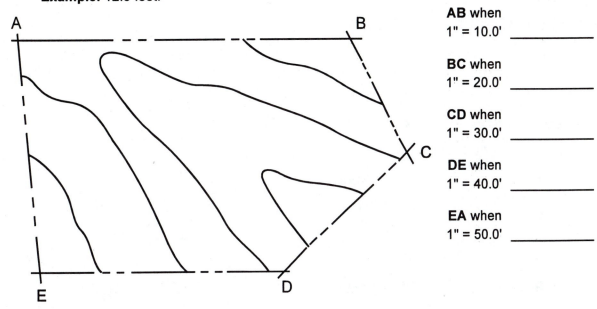

AB when
1" = 10.0' _____

BC when
1" = 20.0' _____

CD when
1" = 30.0' _____

DE when
1" = 40.0' _____

EA when
1" = 50.0' _____

Unit 23 AREA MEASURE

BASIC PRINCIPLES OF AREA MEASURE

Surface or area measure refers to the measurement of an object or part which has height and width, but no thickness. The lengths of all sides of an object must be in the same units before they are multiplied.

The following are formulas of area measure.

Triangle

$A = \dfrac{ab}{2}$

a = altitude

b = base

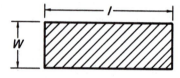

Rectangle

$A = lw$

l = length

w = width

Square

$A = s^2$

s = length of side

Trapezoid

$A = \dfrac{(B + b)\,a}{2}$

B = length of large side

b = length of small side

a = altitude (height)

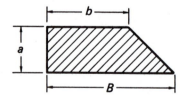

(A = area)

Circle

$A = \pi r^2$

$A = \dfrac{\pi}{4}D^2$

π = 3.141 6

r = radius

D = diameter

$\dfrac{\pi}{4}$ = 0.785 4

PRACTICAL PROBLEMS

Area of Triangles, Squares, Rectangles, and Trapezoids

1. How many square inches are there in a sheet of drafting vellum 18" by 24"?

   _____

2. What is the surface area, in square millimeters, of planes **A** and **B** on this set-up block?

   A

   B

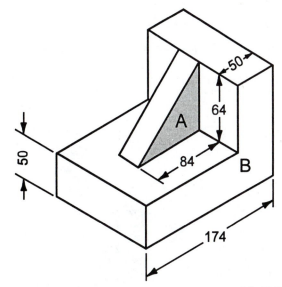

3. How many square inches are there in a square with 7.53-inch sides? Round to the nearest hundredth.

   _____

4. Find, in square feet, the area of the floor of a 19'-0" by 13'-0" blueprint room.

   _____

5. What is the area of this parallelogram in square millimeters?

   _____

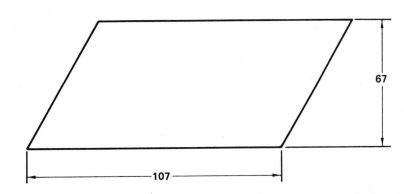

6. Find the floor area as illustrated in this diagram. Express answer in square feet.

_____

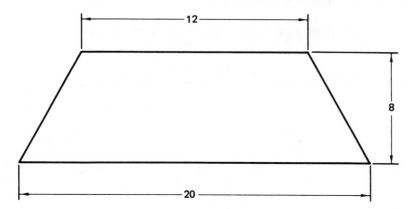

7. What is the area, to the nearest thousandth square inch, of the triangle inscribed in this circle?

_____

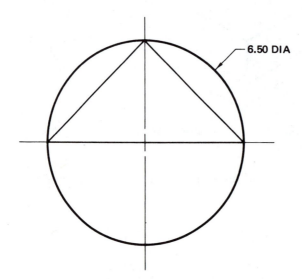

8. What is the difference, in square inches, between the area of a 9" × 12" and a 24" × 36" piece of drafting vellum?

_____

9. Find, in square millimeters, the area of the triangle inscribed in this circle.   _____

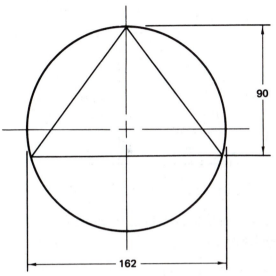

10. What is the area, in square inches, of a drafter's 30–60 degree triangle minus the cutout? Express the answer to the nearest thousandth.   _____

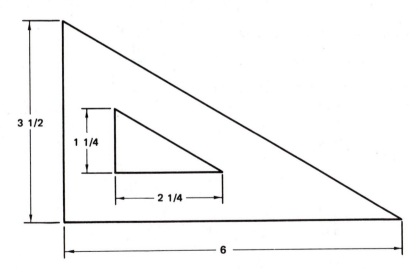

11. Find the weight of a piece of 20-gauge black iron measuring 48" × 96". The weight of 20-gauge black iron is 1.5 lb. per sq. ft.   _____

12. A piece of sheet metal is punched as shown. Find, in square millimeters, the area left after the punching operation. Express answer to the nearest hundredth.

_____

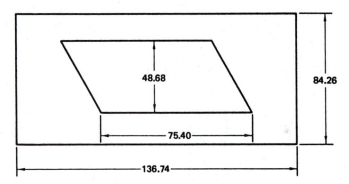

13. The triangular steel plate has a square hole in it. Find, in square inches, the remaining area of the triangular piece.

_____

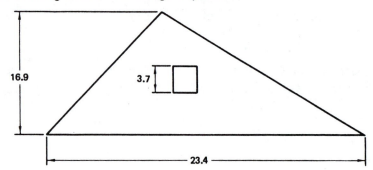

14. What is the area, in square feet, of the side elevation of this house? Do not include the four-foot square window.

_____

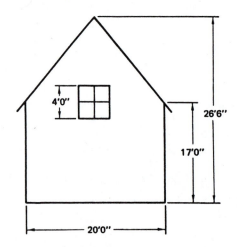

15. How much weight is saved by cutting a rectangular hole in this 22-gauge black iron shim as shown? The dimensions of the hole are in inches. (The weight of 22-gauge black iron is 1.25 lb. per sq. ft.) _____

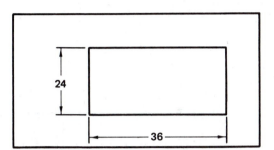

16. What is the total area, in square millimeters, of this transition piece? Express answer to the nearest thousandth. _____

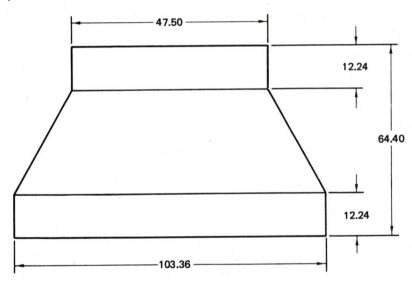

17. A steel block has an angular groove milled through its upper surface. How many square inches less is the top surface area? Round answer to the nearest thousandth. _____

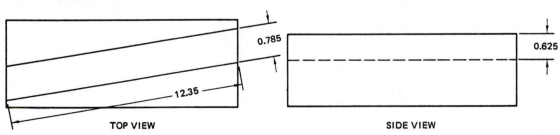

TOP VIEW   SIDE VIEW

18.   Find, in square millimeters, the total area of this template.

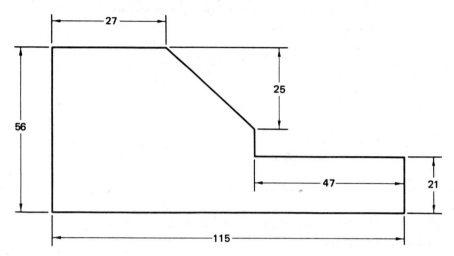

19.   Find, in square inches, the total area of this hopper piece. Express answer to the nearest hundredth.

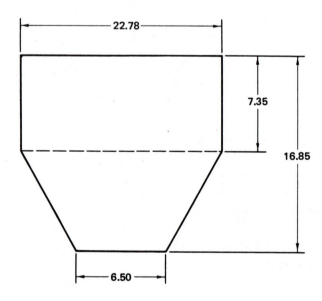

20. Find, in square inches, the amount of waste in this sheet metal piece. Express answer to the nearest hundredth.

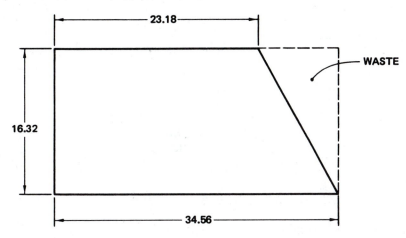

Area of Circles and Parts of Circles

21. Find, to the nearest hundredth inch, the diameter of a circle whose area is 188 square inches.

22. Find, to the nearest thousandth square foot, the area of a circular table top with a diameter of 8 feet.

23. What is the area of the sector removed from this sheet metal disc? Express the answer to the nearest thousandth square inch.

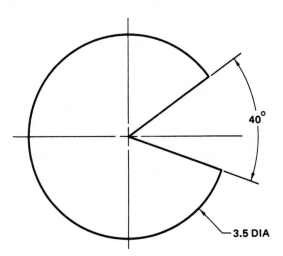

24. What is the area of a washer with an inside diameter of 1¾ inches and an outside diameter of 3⅝ inches? Express the answer to the nearest thousandth square inch.

_____

25. Find, to the nearest hundredth square centimeter, the area of this shim. Hint: Subtract the area of the holes.

_____

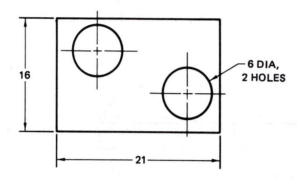

16

6 DIA,
2 HOLES

21

26. Using this stamping, find the area to the nearest hundredth square inch.

_____

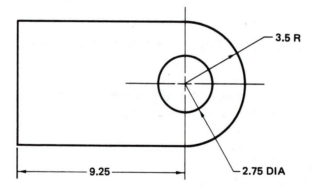

3.5 R

9.25

2.75 DIA

27. A stamping made of 22-gauge tin plate weighs 1.263 lb. per square foot. When the area is 78 square inches, what is the weight? Round answer to the nearest thousandth.

_____

28. Find the total surface area of a right cylinder 4.10 inches in diameter and 8.5 inches high. Express the answer to the nearest hundredth square inch.

_____

29. What is the area of this brass gasket to the nearest thousandth square inch? _____

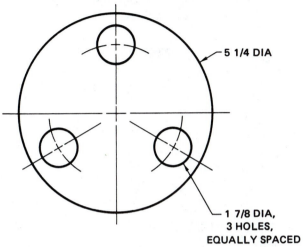

5 1/4 DIA

1 7/8 DIA,
3 HOLES,
EQUALLY SPACED

30. Find, in square millimeters, the area of this circular ring. _____

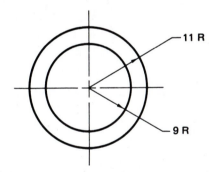

11 R

9 R

31. What is the area of this shim? Express the answer to the nearest thousandth square inch. _____

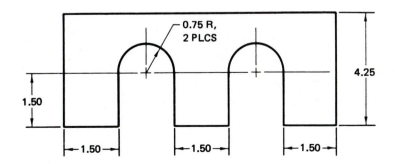

0.75 R,
2 PLCS

4.25

1.50

1.50   1.50   1.50

32. This special cover is made from 30-gauge tin plate which weighs 0.491 lb. per square foot. Find, to the nearest hundredth pound, the weight of the cover. The dimensions on the cover are in inches. _____

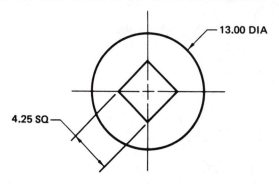

4.25 SQ

13.00 DIA

33. What is the area of this connector link? Express the answer to the nearest hundredth square inch. _____

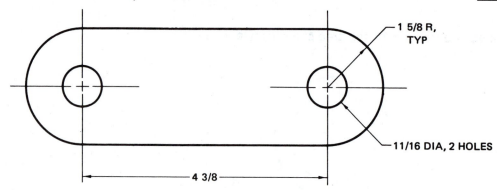

1 5/8 R, TYP

11/16 DIA, 2 HOLES

4 3/8

34. Find the area of the shaded portion of the washer. Express the answer to the nearest thousandth square inch. _____

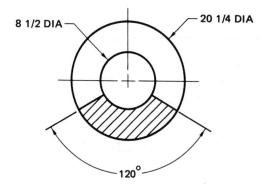

8 1/2 DIA

20 1/4 DIA

120°

35. How many square inches of waste are there in this sheet metal stamping? Express the answer to the nearest thousandth. _____

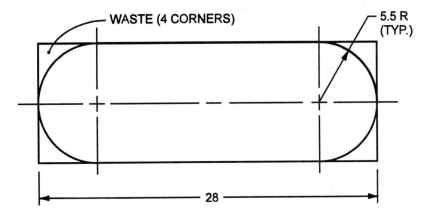

36. Find the area of this circle after the triangular piece is removed. Express the answer to the nearest hundredth square centimeter. _____

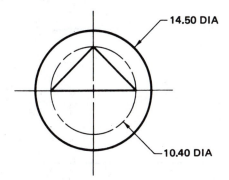

37. What is the lateral surface area of this stepped shaft to the nearest thousandth square centimeter? _____

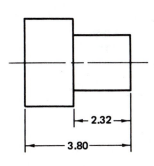
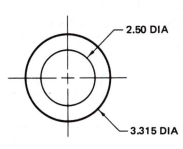

38. Find, to the nearest thousandth square inch, the area of this template. What is the percent of waste for the removal of three semicircular punchings? Express answer to the nearest whole percent.

_____

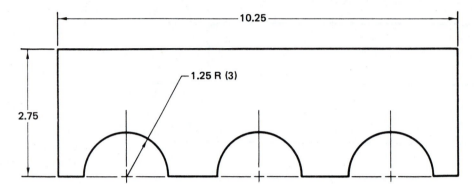

39. What is the area of this shim to the nearest thousandth square decimeter?

_____

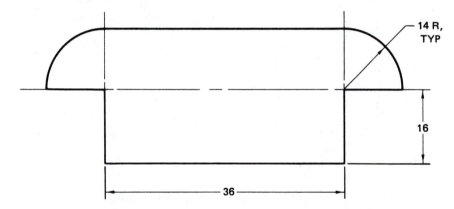

40. What is the total area of 4 templates like the one shown? Express the answer to the nearest thousandth square inch.

_____

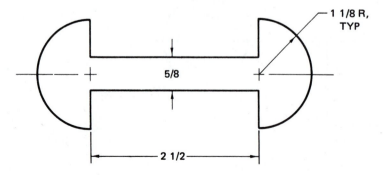

41. Using the CAD drawing presented, determine the area of the cabin plus the area of the deck.

_____

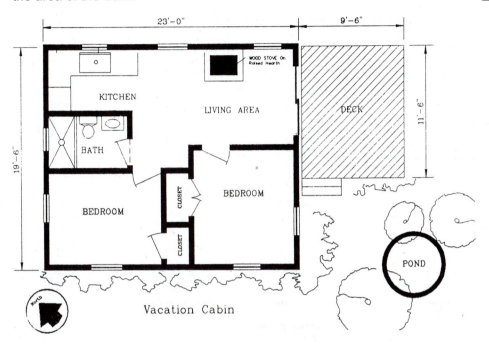

Reprinted with permission from McGrew,
Exploring the Power of AutoCAD, copyright 1990 by Delmar Publishers.

42. Using the CAD drawing illustrated, find the object's area in square inches.

_____

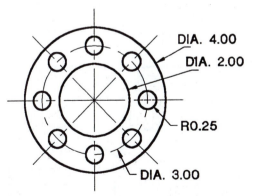

Reprinted with permission from Stellman, Krishnan, and Rhea,
Harnessing AutoCAD, copyright 1993 by Delmar Publishers.

43. Using the CAD drawing shown, find the area of the object in square inches. _____

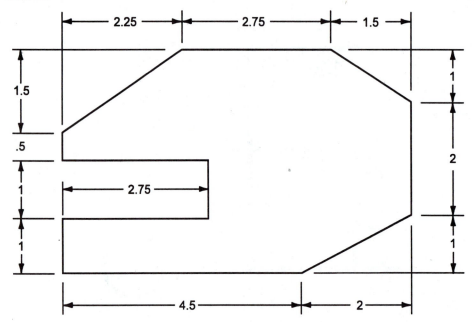

Reprinted with permission from McGrew,
Exploring the Power of AutoCAD, copyright 1990 by Delmar Publishers.

44. Using the CAD drawing of the house, determine the total area consumed by all windows. State your answer in square feet. _____

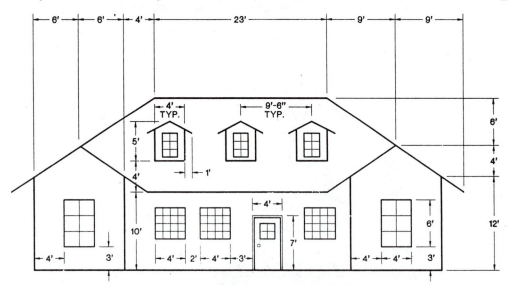

Reprinted with permission from Stellman, Krishnan, and Rhea,
Harnessing AutoCAD, copyright 1993 by Delmar Publishers.

45. Using the CAD drawing given, find the area of the object in square inches. All holes have a ⅝ " diameter. _____

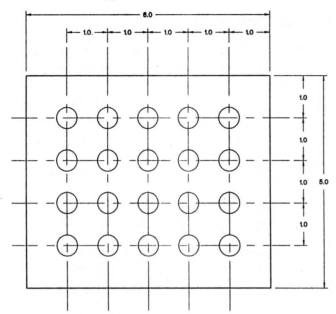

Reprinted with permission from Stellman, Krishnan, and Rhea,
Harnessing AutoCAD, copyright 1993 by Delmar Publishers.

46. Using the CAD drawing of the special gasket, find its area in square inches. _____

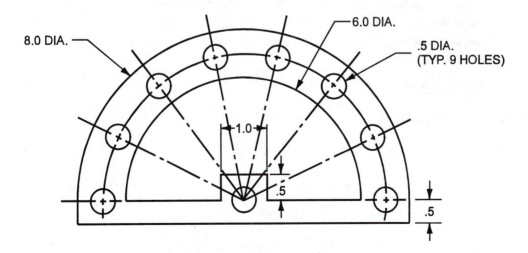

Reprinted with permission from Stellman, Krishnan, and Rhea,
Harnessing AutoCAD, copyright 1993 by Delmar Publishers.

Unit 24 VOLUME MEASURE

BASIC PRINCIPLES OF VOLUME MEASURE

Volume measure is used to determine the space occupied by a body and is usually found by multiplying length by height by width. The following are formulas of volume area.

RECTANGULAR PRISM	$V = L \times H \times W$	L = Length H = Height W = Width
SQUARE PRISM	$V = S^3$	S = Length of side
SQUARE-BASED PYRAMID	$V = \dfrac{Bh}{3}$	B = area of square base h = height of pyramid
CYLINDER	$V = \pi r^2 h$	r = radius of the circular end π = 3.1416 h = height of the cylinder
CONE	$V = \dfrac{\pi r^2 h}{3}$	r = radius of the circular end π = 3.1416 h = height of the cone

In many instances, drafters and engineers are concerned with the weights of shafts and other common machine parts. The weight of a machine part or the combined weight of a product might affect the operation of the product as intended. Thus, volume calculations are extremely important. The volume is multiplied by the weight per unit of volume to determine the total weight.

PRACTICAL PROBLEMS

Volume of Prisms and Pyramids

1. How many cubic inches are there in 4 cubic feet? _____

2. How many cubic feet are there in 9 cubic yards? _____

3. Express 6 cubic yards, 13 cubic feet as cubic feet. _____

4. What is the volume, in cubic inches, of a 1-foot square prism? _____

5. Find, to the nearest cubic foot, the volume of a room 12 ft. 6 in. wide by 20 ft. 6 in. long by 8 ft. 3 in. high.

_____

6. A rectangular solid has sides 8 inches in length and a volume of 640 cubic inches. Find the depth in inches.

_____

7. Find, in cubic millimeters, the volume of this step block.

_____

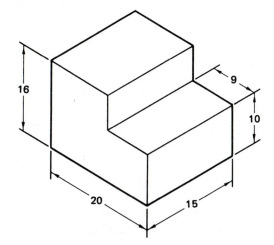

8. What is the volume of this rectangular core box in cubic inches?

_____

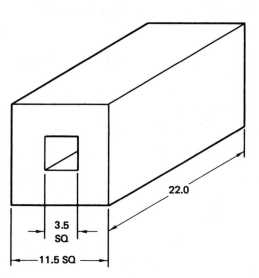

9.   Find, in cubic inches, the volume of this notch block. Express the answer
     to the nearest thousandth.   _____

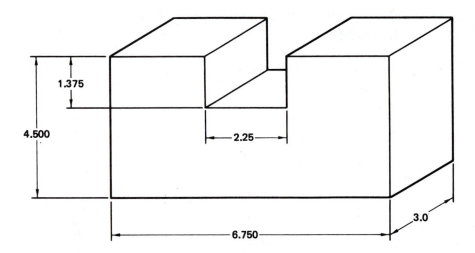

10. A rectangular water storage tank measures 57 in. by 74 in. by 42 in.
     Find, to the nearest hundredth, the gallon capacity of the tank.   _____

11. A rectangular tank with height 38.5 in. and length 25.25 in. has a 57.0-
     gallon capacity. What is the width of the tank to the nearest hundredth
     inch?   _____

12. A piece of wood measures 16 in. by 8 in. and contains 512 cubic inches.
     How many inches thick is the wood?   _____

13. A specific type of steel weighs 0.283 pounds per cubic inch. Find the
     weight of a square bar with 3-inch sides and a length of 7.5 inches.
     Express the answer to the nearest hundredth.   _____

14. Find, to the nearest hundredth cubic yard, the volume of this square-based pyramid. _____

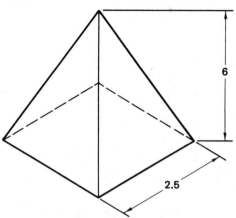

15. Determine the number of cubic yards of concrete in a pyramid with a square base measuring 8 yards on each side and an altitude of 42 feet. Express answer to the nearest hundredth. _____

16. Find the difference, in cubic yards, between a cube with 3-foot sides and a cube with 8-foot sides. Express the answer to the nearest hundredth. _____

17. A piece of steel 2 feet long and 2 feet wide weighs 490 pounds. Steel weighs 490 pounds per cubic foot. What is the height of this piece of steel to the nearest $\frac{1}{4}$ foot? _____

18. What is the volume, in cubic inches, of a piece of wood $\frac{3}{4}$ inch by 3 inches by 12 inches? _____

19. Find, in cubic feet, the volume of this wall. _____

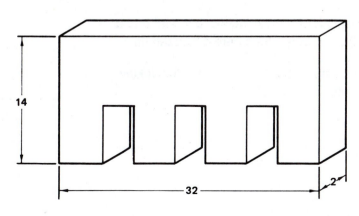

NOTE: ALL NOTCHES ARE 6 ft. × 3 ft.

20. Determine the volume of this stop block. Express the answer to the nearest thousandth cubic inch.

_____

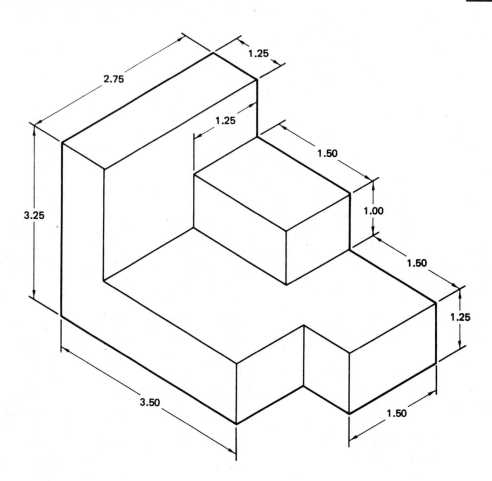

Volume of Cylinders and Cones

21. Find, in inches, the volume of a cylinder that is 8 inches long and has a diameter of 1.5 inches. Express the answer to the nearest thousandth.

_____

22. A cone is 7.5 feet high and has a diameter of 2.5 feet. What is the volume of the cone to the nearest thousandth cubic foot?

_____

23. Find the volume of this centering pin to the nearest thousandth cubic inch.

_____

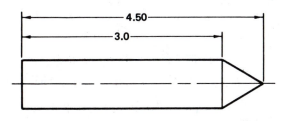

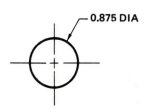

24. What is the volume, in cubic feet, of a cylinder 15 inches in diameter and 14 feet long? Express the answer to the nearest hundredth.

_____

25. Find, in cubic yards, the volume of a cone 25 feet in diameter and 18 feet high. Express the answer to the nearest hundredth.

_____

26. A 2-inch square hole is machined in this cylinder. Find, to the nearest thousandth cubic inch, the volume after machining the square hole.

_____

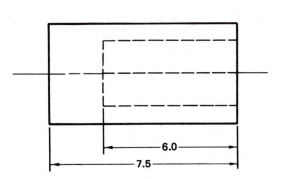

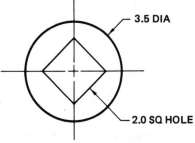

27. A cylinder-shaped oil storage tank is 8 ft. in diameter and 15 ft. high. What is the capacity to the nearest hundredth gallon?

_____

28. What is the volume of this holding block to the nearest hundredth cubic centimeter? _____

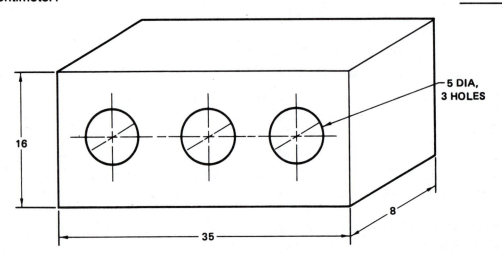

5 DIA,
3 HOLES

16

35

8

29. Find, to the nearest thousandth cubic inch, the volume of this centering pin. _____

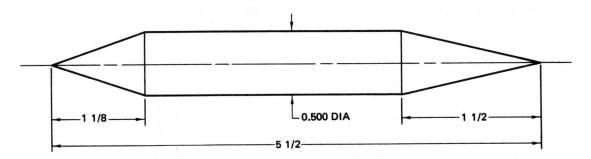

1 1/8

0.500 DIA

1 1/2

5 1/2

30. Find, to the nearest hundredth pound, the weight of this bronze bushing. Bronze weighs 3.2 pounds per cubic inch. All dimensions on the bronze bushing are in inches. _____

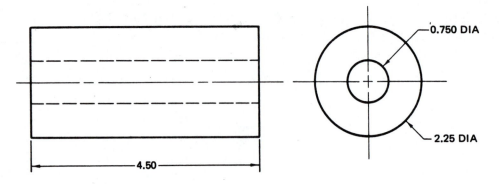

4.50

0.750 DIA

2.25 DIA

31. Find the gallon capacity of a cylinder-shaped oil drainage pan 30 inches in diameter and 8 inches high. Express the answer to the nearest thousandth.   _____

32. A flat-bottomed hole is milled in this retainer cap. The hole is 7 cm in diameter and 20 cm deep. What is the volume of the cap after the milling operation? Express the answer to the nearest thousandth cubic centimeter.   _____

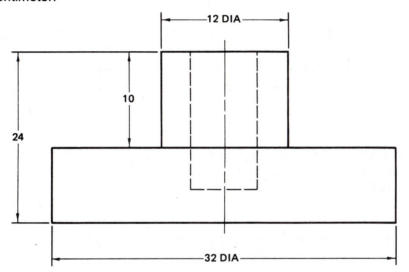

33. Find the volume of a cone 36 mm in diameter and 19 mm in length. Express the answer to the nearest thousandth cubic millimeter.   _____

34. What is the volume of the end portion of this lathe live center to the nearest thousandth cubic inch? The "end portion" includes the conical and cylindrical portions only.   _____

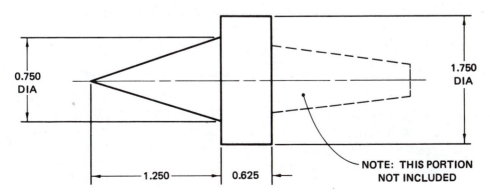

35. Find, to the nearest hundredth pound, the weight of this brass coupling. Brass weighs 3.0 pounds per cubic inch. All dimensions on the brass coupling are in inches.

_____

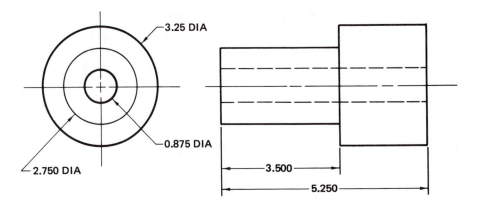

36. A steel collar has an inside diameter of 1 ½ in., an outside diameter of 3 ¾ in., and a length of 2 ⁷⁄₁₆ in. Find, to the nearest hundredth pound, the weight of the collar. The weight of steel is 0.283 pounds per cubic inch.

_____

37. What is the volume of this special bearing cap to the nearest thousandth cubic inch?

_____

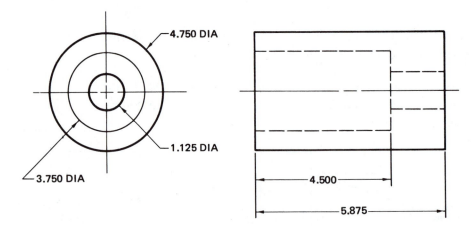

38. A 6-inch long piece of bar stock is turned on a lathe from a 5 ½-inch diameter to a 3 ¼-inch diameter. The bar stock remains cylindrical after the machining operation. How many cubic inches of metal are removed? Express the answer to the nearest hundredth.

_____

39. Find, in pounds, the weight of a piece of bar stock 1 ½ " in diameter and
     8 ⅞ " long. Use 0.283 pounds per cubic inch for the weight of the stock.
     Express the answer to the nearest hundredth.   _____

40. Find, to the nearest thousandth cubic inch, the volume of this hold down
     plate.   _____

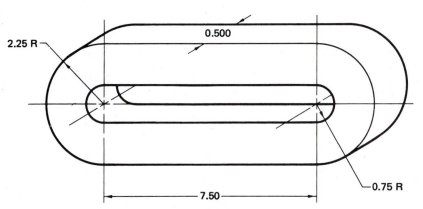

41. Using the CAD drawing of the face plate, find its volume in cubic inches.   _____

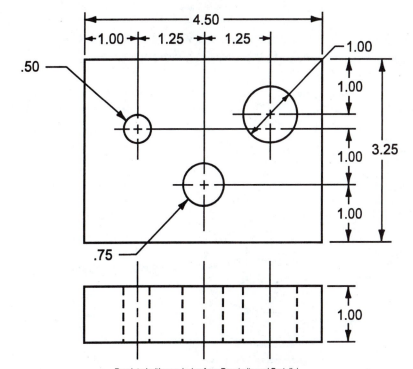

Reprinted with permission from Resetarits and Bertolini,
Using CADKEY Light, copyright 1992 by Delmar Publishers.

42. Using the CAD drawing presented, find the weight of the object if the stock weighs 0.246 pounds per cubic inch. Express your answer to the nearest hundredth.

_____

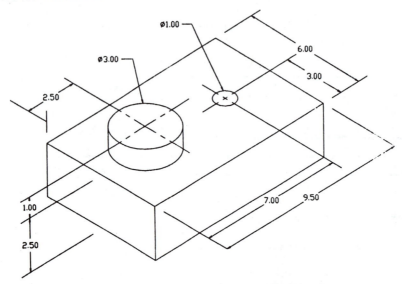

Reprinted with permission from Resetarits and Bertolini,
Using CADKEY Light, copyright 1992 by Delmar Publishers.

43. Using the CAD drawing illustrated, determine the volume of the object in cubic inches. Express your answer to the nearest hundredth.

_____

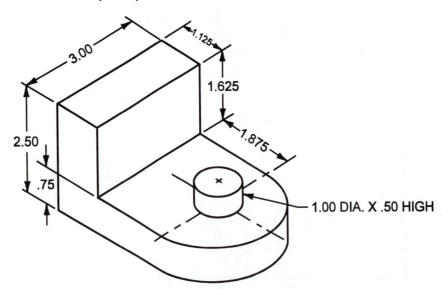

Reprinted with permission from McGrew,
Exploring the Power of AutoCAD, copyright 1990 by Delmar Publishers.

44. Using the CAD drawing shown, find the weight of the object if the stock weighs 0.312 pounds per cubic inch. Express your answer to the nearest hundredth. _____

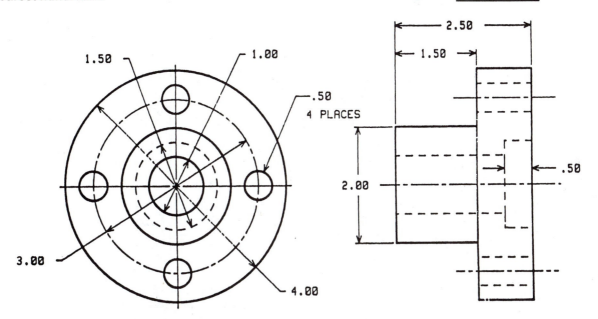

Reprinted with permission from Resetarits and Bertolini,
Using CADKEY Light, copyright 1992 by Delmar Publishers.

45. Using the CAD drawing provided, determine the volume of the object in cubic inches. Express your answer to the nearest hundredth. _____

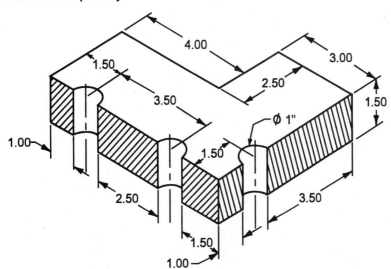

Reprinted with permission from Stellman, Krishnan, and Rhea,
Harnessing AutoCAD, copyright 1993 by Delmar Publishers.

46.   Using the CAD drawing illustrated, find the weight of the object if the stock weighs 0.472 pounds per cubic inch. Express your answer to the nearest hundredth.

_____

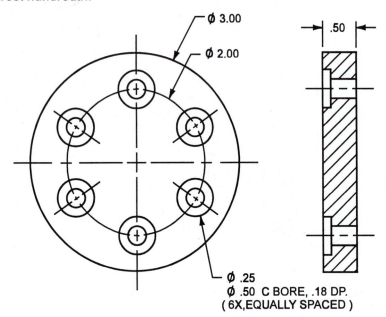

Ø 3.00

Ø 2.00

.50

Ø .25
Ø .50 C BORE, .18 DP.
(6X,EQUALLY SPACED)

Unit 25 EQUIVALENT MEASUREMENT UNITS AND CONVERSION

BASIC PRINCIPLES OF EQUIVALENT MEASUREMENT UNITS AND CONVERSION

The use of metric measure is increasing in the manufacture of products. Machines, tooling, and gauges in a factory are often designed for one system of measurement. Architectural and engineering drawings that are dimensioned in the other system must often be changed.

The metric system is easy to understand if you keep in mind that it is based on multiples of tens, the same as our monetary system. The prefix of each unit tells what fraction or multiple of a meter is being used.

METRIC LENGTH

milli	=	1/1000 of a meter	deci	=	1/10 of a meter
centi	=	1/100 of a meter	kilo	=	1000 meters

TABLE OF LINEAR EQUIVALENTS

1 inch	=	0.0254 meter	39.37 inches	=	1 meter
1 inch	=	0.254 decimeter	3.937 inches	=	1 decimeter
1 inch	=	2.54 centimeters	0.394 inches	=	1 centimeter
1 inch	=	25.40 millimeters	0.039 inches	=	1 millimeter
		1 foot	=	0.3048 meters	
		1 yard	=	0.9144 meters	

TABLE OF AREA EQUIVALENTS

1 square inch	=	6.4516 cm^2	1,550 square inches	=	1 m^2
1 square foot	=	0.0929 m^2	15.50 square inches	=	1 dm^2
1 square yard	=	0.836 m^2	0.155 square inch	=	1 cm^2
			0.00155 square inch	=	1 mm^2

TABLE OF VOLUME EQUIVALENTS

1 cubic inch	=	16.387 cm^3	61,023.377 cubic inches	=	1 m^3
1 cubic foot	=	0.0283 m^3	61.023 cubic inches	=	1 dm^3
1 cubic yard	=	0.7646 m^3	0.061023 cubic inch	=	1 cm^3
			0.000059 cubic inch	=	1 mm^3

PRACTICAL PROBLEMS

Metric-English Linear Equivalents

Round answers to the nearest thousandth.

1. Express 6" as millimeters. _____

2. Express 118 mm as inches. _____

3. Express 16 ½ inches as centimeters. _____

4. Express 4.750" as millimeters. _____

5. Express 14 feet as meters. _____

6. Express 7 meters as feet. _____

7. Express 396.35 centimeters as inches. _____

8. Express 17 meters as yards. _____

9. How many inches equal 17 decimeters? _____

10. How many inches equal 39 centimeters? _____

11. All dimensions on this machined plate are in inches. Express, in millimeters, dimensions **A, B, C,** and **D** on this machined plate.

A _____

B _____

C _____

D _____

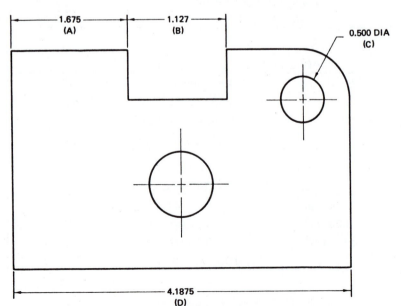

12. What is the diameter, in millimeters, of a 16-inch diameter circle? _____

13. A 72-inch line is subdivided into 8 equal parts. Express the length of each subdivision in millimeters. _____

14. In this step block, find the missing dimension. All dimensions on the step block are in inches. Express the answer in centimeters. _____

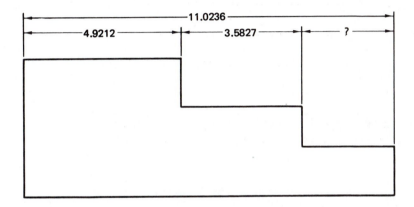

15. A ⁵⁄₁₆" diameter piece of copper round stock is 315 centimeters long. How many 7-inch studs can be cut from it? _____

16. The dimensions on this large washer are in millimeters. Determine the wall thickness, in inches, of the large washer. _____

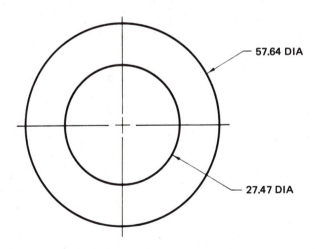

57.64 DIA

27.47 DIA

17. A certain collar has a wall thickness of 11.65 mm. The outside diameter is 57.64 mm. Find the inside diameter in inches. _____

18.   The dimensions on this drilled hole are in millimeters.

    a.   Express the diameter of this drilled hole in inches.   _____

    b.   Express the depth of this drilled hole in inches.   _____

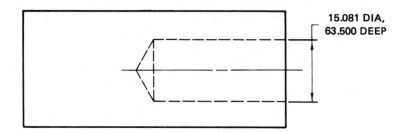

15.081 DIA,
63.500 DEEP

19.   Five line lengths are combined to form one line. They are 2.54 cm, 5.08 cm, 22.225 mm, 6.350 mm, and 0.9525 cm. What is the length of this line in inches?   _____

20.   Determine dimensions **A** and **B** in this fixture. The dimensions on the fixture are in inches. Express each answer in decimeters.   A   _____

    B   _____

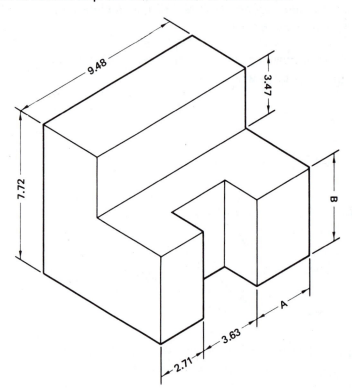

Metric-English Area Equivalents

Note: Answers may vary, depending on approximation (rounding) of numbers.

21. Express 12 inches as centimeters. _____

22. Express 939.8 millimeters as inches. _____

23. Express 19 square decimeters as square inches. _____

24. Express 12.245 square inches to the nearest square centimeter. _____

25. Find, in square centimeters, the area of a circle with a diameter of 1 $\frac{9}{16}$ inches. Express the answer to the nearest hundredth. _____

26. A mechanical drafter must find the area, in square centimeters, of a rectangle 2 $\frac{1}{16}$ inches long and 1 $\frac{3}{8}$ inches wide. State the answer to the nearest hundredth. _____

27. All dimensions on this triangle are in inches. Find, in square millimeters, the area of this triangle. Express the answer to the nearest hundredth. _____

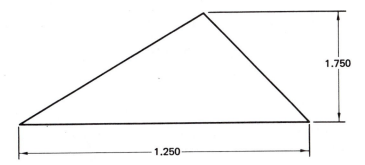

28. Find the area, in square inches, of a circle with a diameter of 254.00 millimeters. State the answer to the nearest hundredth. _____

29. A drawing storage room in an engineering firm measures 22 feet by 32 feet. What is the area of the floor to the nearest hundredth square meter? _____

30. The floor area of a blueprint room is 60 square meters. What is the area to the nearest hundredth square yard? _____

31. The floor area of a microfilming room in a small company is 16.7 square meters. What is the area to the nearest hundredth of a square yard? _____

32. All dimensions on this trapezoid are in inches. Fine the area, in square decimeters, of this trapezoid. Express the answer to the nearest thousandth.

_____

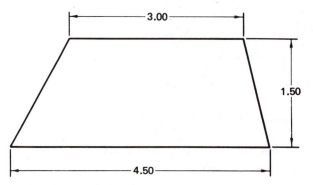

33. All dimensions on this rectangle are in millimeters. Find, to the nearest thousandth square inch, the area of the rectangle.

_____

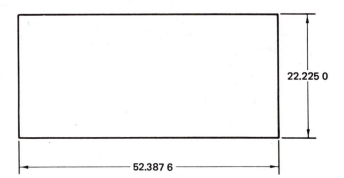

34. Find the area, in square meters, of a circular table with a diameter of 4 feet. Express the answer to the nearest thousandth.

_____

35. All dimensions on this circular ring are in decimeters. Find, to the nearest thousandth square inch, the area of the circular ring.

_____

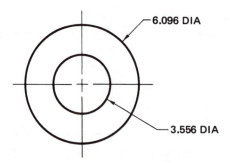

36. All dimensions on this parallelogram are in meters. Find, to the nearest hundredth square inch, the area of the parallelogram. _____

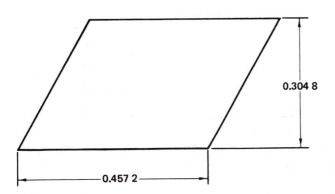

37. A square piece of galvanized sheet metal has an area of 67.59 square meters. Find the area to the nearest thousandth square yard. _____

38. This template has an area of 43.7 square inches. What is the area to the nearest hundredth square centimeter? _____

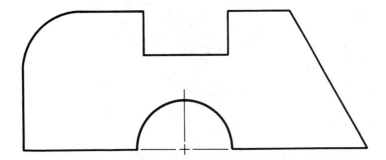

39. This shim has an area of 27.93 square meters. What is the area to the nearest thousandth square foot? _____

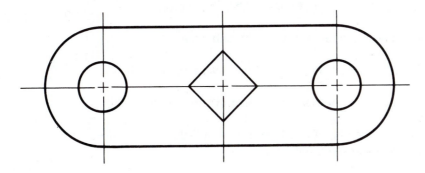

40. This template has an area of 1,806.40 square millimeters. What is the area to the nearest hundredth square inch?

_____

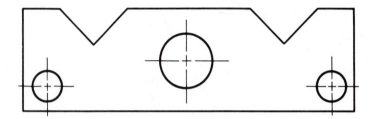

41. All dimensions on this template are in centimeters. Find, to the nearest hundredth square inch, the area of the template.

_____

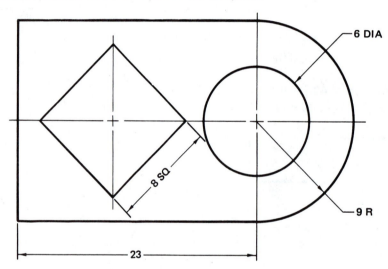

Metric-English Volume Equivalents

42. Express 5,184 cubic inches to the nearest thousandth cubic meter.

_____

43. Express 18 cubic yards to the nearest hundredth cubic meter.

_____

44. Express 117 cubic feet to the nearest hundredth cubic meter.

_____

45. Express 15.4 cubic decimeters to the nearest hundredth cubic inch.

_____

46. Express 5.429 cubic inches to the nearest thousandth cubic centimeter.

_____

47. Find the volume, in cubic inches, of a cubical box with 3-decimeter sides. Express the answer to the nearest hundredth.

_____

48. What is the total volume of this gauge? State answer to the nearest hundredth cubic inch. All dimensions on the gauge are in centimeters. _____

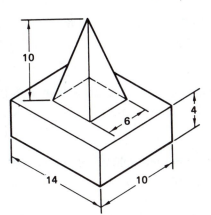

49. A block of steel is 3 meters long, 3 meters wide, and 1 meter high. Find, to the nearest hundredth cubic yard, the volume of the block. _____

50. Find, in cubic decimeters, the volume of this steel collar. Express the answer to the nearest ten-thousandth cubic decimeter. All dimensions on the steel collar are in inches. _____

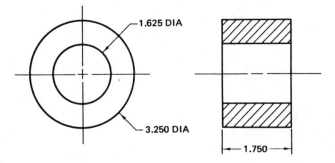

51. The dimensions of this core box are in yards. Find, to the nearest hundredth cubic meter, the volume of the box. _____

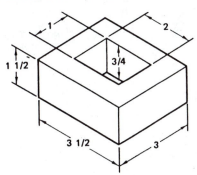

52. What is the volume, in cubic decimeters, of a cylinder with a radius of 0.75 ft. and a length of 4 ft.? Express the answer to the nearest hundredth. _____

53. The height of a square pyramid is 0.95 m and the base measures 0.46 m on each side. Find, to the nearest thousandth cubic foot, the volume of the pyramid. _____

54. Determine the volume, in cubic centimeters, of a cone with a 6-inch diameter base and an altitude of 14 inches. State the answer to the nearest hundredth. _____

55. Find, to the nearest hundredth cubic inch, the volume of this spacer. All dimensions on the spacer are in centimeters. _____

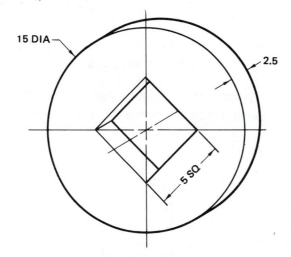

56. All dimensions on the holding block are in meters. Find, to the nearest hundredth cubic yard, the volume of the block. _____

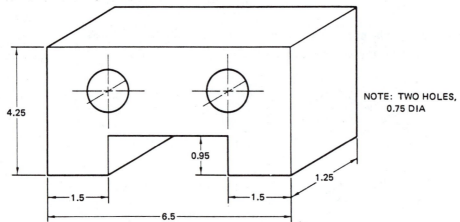

57. Find the volume, in cubic decimeters, of this gasket. Express the answer to the nearest hundredth. All dimensions on the gasket are in inches. _____

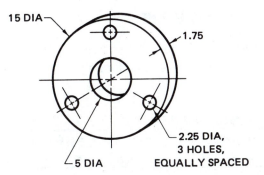

15 DIA

1.75

2.25 DIA,
3 HOLES,
EQUALLY SPACED

5 DIA

58. What is the volume of a drafting room 50 ft. by 24 ft. by 9 ft.? State the answer to the nearest hundredth cubic meter. _____

59. A cone-shaped pile of sand is 18 ft. high and has a base diameter of 25 ft. Find, to the nearest hundredth cubic meter, the volume of the pile. _____

60. What is the volume, in cubic millimeters, of a rectangular mylar film eraser that measures 0.375 in. by 1.5 in. by 0.750 in.? Express the answer to the nearest hundredth cubic millimeter. _____

61. Find, to the nearest hundredth cubic inch, the volume of this step block. The levels are 4-cm, 7-cm, and 13-cm squares and are each 3 cm high. A 1.80-cm diameter hole passes through all levels. _____

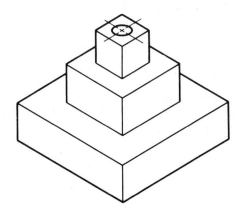

Unit 26 ANGULAR MEASURE

BASIC PRINCIPLES OF ANGULAR MEASURE

All types of drawings contain angular features. The size of an opening or corner on a drawing must be dimensioned. Degrees, minutes, and seconds are used for angular measurement in the English and metric systems.

ANGULAR MEASURE				
1 circumference (circle)	= 360°	1 degree (1°)	=	60 minutes (60')
1/4 circle	= 90°	1 minute (1')	=	60 seconds (60")

To change degrees to minutes, multiply the number of degrees by 60 and then express the product in terms of minutes.

Example: Change 5 degrees (5°) to minutes.

$5 \times 60 = 300$
Answer: $5° = 300'$

To change minutes to seconds, multiply the number of minutes by 60 and then express the product in terms of seconds.

Example: Change 9 minutes (9') to seconds.

$9 \times 60 = 540$
Answer: $9' = 540"$

PRACTICAL PROBLEMS

1. How many degrees are there in a semicircle? _____

2. How many 30-degree angles are there in a circle? _____

3. How many degrees are there in 3 right angles? _____

4. How many degrees are there in $\frac{1}{5}$ of a circle? _____

5. How many degrees and minutes are there in angle **X**?   _____

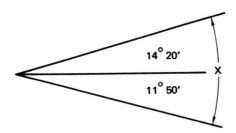

6. How many degrees are there in angle **X**?   _____

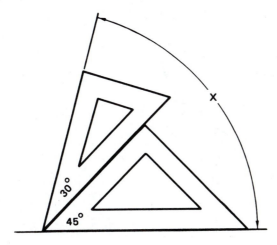

7. One angle measure 8° 15'. What is the total number of degrees and minutes in 7 of these angles?   _____

8. How many degrees are there in angle **X**?   _____

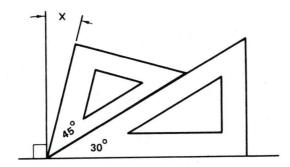

9. Find the number of degrees between the centers of two consecutive holes on this base plate. _____

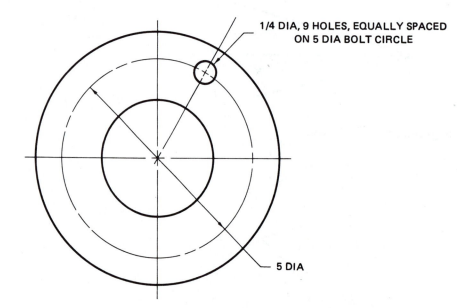

1/4 DIA, 9 HOLES, EQUALLY SPACED
ON 5 DIA BOLT CIRCLE

5 DIA

10. Find, in degrees and minutes, angle **X**. _____

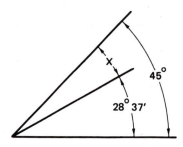

45°

28° 37′

11. How many degrees are there in a straight angle? _____

12. Express 720 minutes as degrees. _____

13. What is the total number of degrees in 840 minutes and 10,800 seconds? _____

14. Find, in degrees and minutes, angle **X**. _____

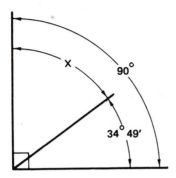

15. Find, in degrees, minutes, and seconds, angle **Y**. _____

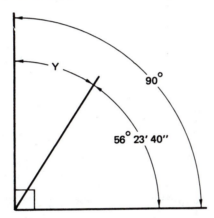

16. Angle **R** plus angle **S** equals 180°. Angle **S** measures 67° 13'. Find, in degrees and minutes, angle **R**. _____

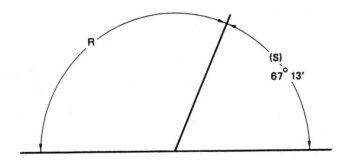

17. Two angles measure 180°. The first angle is 114° 28' 37". Find the number of degrees, minutes, and seconds in the second angle. _____

18. On the circle, center lines **RS** and **TU** form right angles. Find the value of angles **C, E,** and **G**.

C _____

E _____

G _____

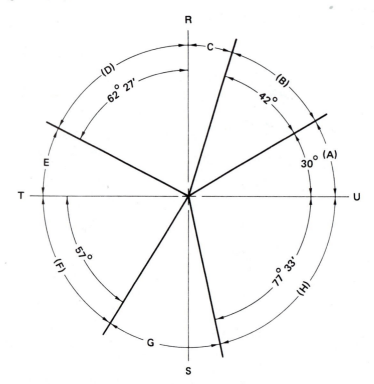

19. Using a protractor, find the number of degrees in the labeled angles on this template.

A _____

B _____

C _____

D _____

E _____

F _____

20. An angle measuring 26° 42' 36" is divided into 4 equal parts. Find the number of degrees, minutes, and seconds in each part. _____

21. The interior angles on the site plan total 360°. Find angle **D**. _____

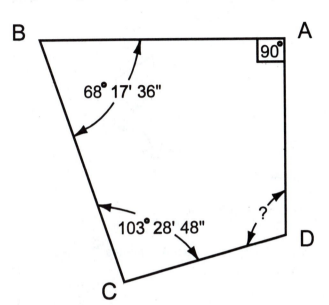

22. A regular pentagon has five equal sides and five equal interior angles. Find the angle indicated. _____

23. Using the CAD drawing shown, determine angles **A–D**.

A _____

B _____

C _____

D _____

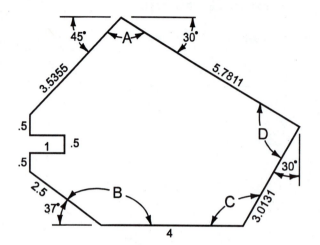

24. Using the CAD drawing illustrated, find the values for angles **C** and **E** if angle **A** is 37° 46' and angle **D** is 27° 38'.

C _____

E _____

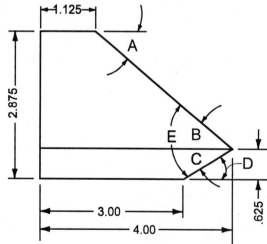

25. Using the CAD drawing presented, find angles **C** and **D** if angle **A** is 22°
    18', and angle **B** is 48° 44'.

    C _____

    D _____

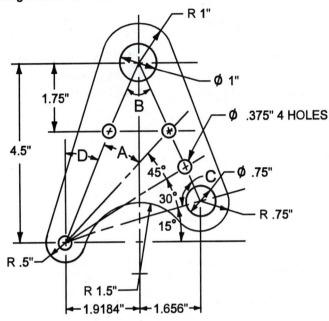

Reprinted with permission from Stellman, Krishnan, and Rhea,
Harnessing AutoCAD, copyright 1993 by Delmar Publishers.

26. Using the CAD drawing below, determine angles **C** and **F**, if **A** is 43° 37',
    B is 33° 41', **D** is 25° 16', and **E** is 66° 21'.

    C _____

    F _____

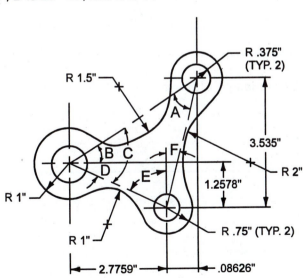

Reprinted with permission from Stellman, Krishnan, and Rhea,
Harnessing AutoCAD, copyright 1993 by Delmar Publishers.

Unit 27 SCALED MEASUREMENT

BASIC PRINCIPLES OF SCALED MEASUREMENT

The drafter must often draw parts that are too large or too small to be drawn full scale. The drafter must then choose and convert the actual measurements to a proper scale. If a part is large, the dimensions for the drawing are reduced. If a part is small, the dimensions for the drawing are increased to a scale that permits easy reading.

PRACTICAL PROBLEMS

Use full scale for questions 1–8 to the degree of accuracy indicated for each problem.

1. Measure and record the full-scale length of these lines to the nearest $\frac{1}{4}$".

    a ├───┤

    b ├──────────────────────────────────┤

    a _____

    b _____

2. Measure and record the full-scale length of these lines to the nearest $\frac{1}{8}$".

    a ├──────────────────┤

    b ├──┤

    a _____

    b _____

3. Measure and record the full-scale length of these lines to the nearest $\frac{1}{16}$".

    a ├────────────────────────────┤

    b ├─────────────────────────────────────┤

    a _____

    b _____

4. Measure and record the full-scale length of these lines to the nearest mm.

    a ├──────────────────────────┤

    b ├─────────────────────────────────────┤

    a _____

    b _____

5. Measure and record the full-scale length of these lines to the nearest ¹⁄₆₄ ".

   a ├──────────────┤   a _____

   b ├──────────────────────────────┤   b _____

6. Measure and record the full-scale length of these lines to the nearest cm.

   a ├────────────────────────────────┤   a _____

   b ├─────────────────────┤   b _____

7. Measure and record the full-scale length of these lines to the nearest ¹⁄₃₂ ".

   a ├──────────────────────────────────┤

   b ├─────────────────────┤   a _____

                                          b _____

8. Measure and record the full-scale length of these lines to the nearest mm.

   a ├────────────────────────────────┤   a _____

   b ├──────────────────────┤   b _____

9. On this drawing, ¼ inch represents 1 inch on the part. Measure each
   labeled dimension, to the nearest ¼ ", and determine its length on the part.   A _____

   B _____

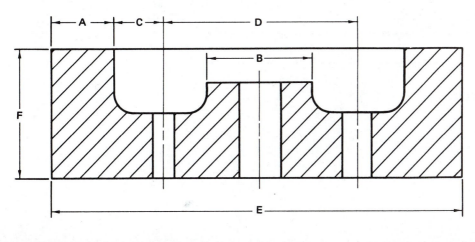

   C _____

   D _____

   E _____

   ·F _____

SCALE $\frac{1''}{4}$ = 1"

10. Measure each labeled dimension, to the nearest ⅛", and determine its length on the part. The scale is ½" = 1".

A _____

B _____

C _____

D _____

E _____

F _____

11. Measure this full-scale drawing and record the lengths to the nearest millimeter.

A _____

B _____

C _____

D _____

E _____

F _____

G _____

H _____

12. On this drawing, ⅜" represents 1" on the part. Measure these labeled dimensions and determine the lengths on the part to the nearest ⅛".

A _____

B _____

C _____

D _____

E _____

F _____

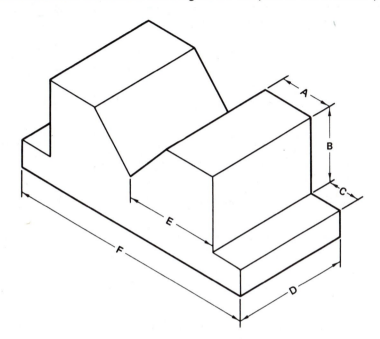

13. This drawing was made with the scale ¾" = 1". What are these dimensions on the part to the nearest ⅛"?

A _____

B _____

C _____

D _____

E _____

F _____

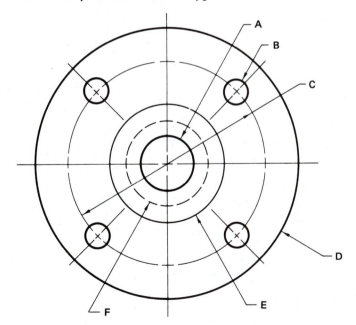

14.   Measure all lengths in centimeters.

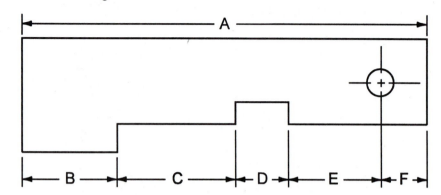

A	_____
B	_____
C	_____
D	_____
E	_____
F	_____

15.   Measure all sides of the plot plan using the scale indicated.

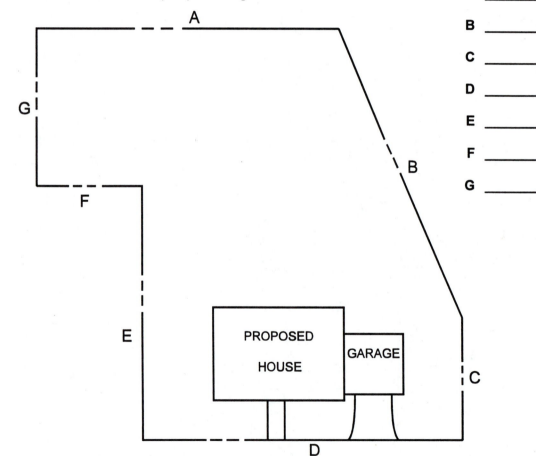

SCALE: 1" = 100.0'

A	_____
B	_____
C	_____
D	_____
E	_____
F	_____
G	_____

16. Using the scale indicated, measure and record the dimensions indicated.

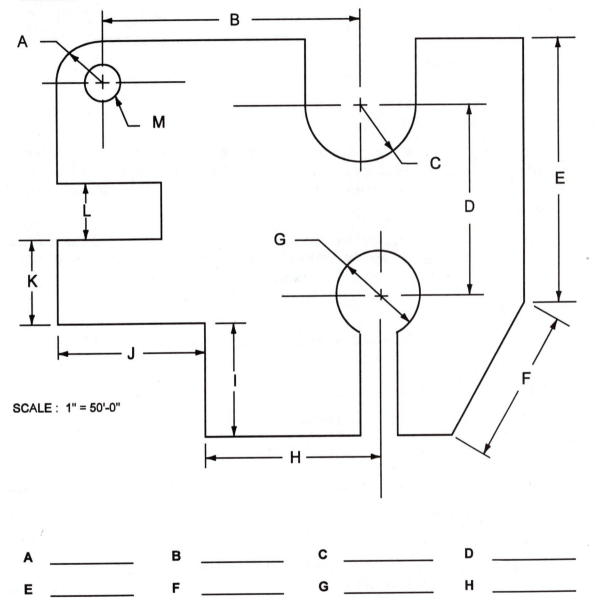

SCALE : 1" = 50'-0"

A _____   B _____   C _____   D _____

E _____   F _____   G _____   H _____

I _____   J _____   K _____   L _____

M _____

Ratio and Proportion

Unit 28 RATIO

BASIC PRINCIPLES OF RATIO

Full-scale drawings are not always practical. Computer plots of large objects must be reduced in order to be drawn on paper. Drawings of small objects must be enlarged to see the details clearly. Drawings are made to different scales, such as quarter-size, half-size, or double-size, which are actually ratios.

A *ratio* is a comparison of two numbers or quantities that are similar. Just like fractions, ratios should be reduced to their simplest form. The ratio 10:5 (read as 10 to 5) is reduced to 2:1 by dividing both numbers by 5. The ratio 3:7 can be also written in fractional form.

Example: $3:7 = \frac{3}{7}$

An *inverse ratio* (reciprocal) is the reverse of the original ratio. The inverse ratio of 3:7 is 7:3 or $\frac{7}{3}$.

PRACTICAL PROBLEMS

Express the ratios of these dimensions in simplest form.

1. 3 cm to 16 mm   _____

2. 6 in. to 2 ft.   _____

3. 4 ft. to 8 in.   _____

4. 2 yd. to 3 in.   _____

5. Gear *A* has 60 teeth and gear *B* has 12 teeth. What is the ratio of gear *A* to gear *B*?   _____

6. In a drawing, a scale of 1" = 1'-0" is used. What is the ratio of the size of the drawing to the actual object?   _____

7. In a drawing, a scale of $\frac{3}{4}$" = 1'-0" is used. What is the ratio of the size of the drawing to the actual object?   _____

8. Find the ratio of the circumference of the smaller circle to the circumference of the larger one.

_____

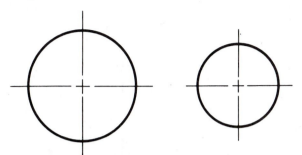

CIRCUMFERENCE = 62.832"   CIRCUMFERENCE = 47.124"

9. The scale ⅜" = 1'-0" is used in a drawing. What is the ratio of the actual object to the drawing?

_____

10. Measure dimensions **A, B,** and **C**. Using the scale 1:3, find, in inches, the actual lengths.

A _____

B _____

C _____

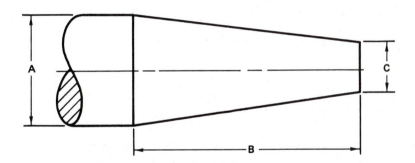

11. When the scale ¼" = 1" is used, what is the ratio of actual object to the drawing being made?

_____

12. It takes an experienced CAD drafter 16 hours to complete a set of working drawings. Another CAD drafter takes 96 hours to complete a similar set of working drawings. What is the ratio of the first CAD drafter's time to the second CAD drafter's time?

_____

13. Measure dimensions **A, B, C,** and **D**. Using the scale 2:1, find, in inches, the actual length of the dimensions.

A _____

B _____

C _____

D _____

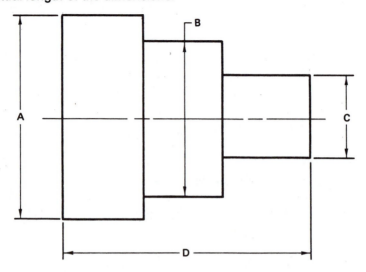

14. The gear ratio of the teeth on gear *A* to gear *B* is 5:1. How many times faster does gear *B* move than gear *A*?

_____

15. When the scale $\frac{1}{2}$ " = 1'-0" is used, what is the ratio of the actual object to the drawing being made?

_____

16. Pulley *B* turns 7 times for every turn of pulley *A*. What is the ratio of the diameter of *A* to the diameter of *B*?

_____

17. Measure the dimensions (sides and hole size) on the shim illustrated. Using the scale 3:1, find, in inches, the actual length of the dimensions. Use a decimal scale and express the answer in hundredths.

A _____

B _____

C _____

D _____

E _____

F _____

G _____

H _____

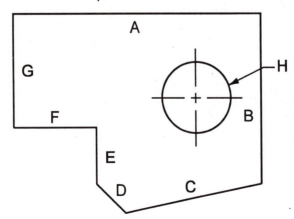

18. What is the inverse ratio of gear *A* with 72 teeth to gear *B* having 12 teeth?

_____

19. What is the ratio of drawing to object size when the scale $\frac{1}{4}$" = 1'-0" is being used?

_____

20. CAD drafter *A* generates 24 architectural symbols in 12 hours. CAD operator *B* generates 24 symbols in 8 hours. What is the ratio of CAD drafter *A*'s time to CAD drafter *B*'s time?

_____

Unit 29 **PROPORTION**

BASIC PRINCIPLES OF PROPORTION

Proportion is the equality of two ratios.

$$3:2 = 9:6$$

There are two basic methods used to solve proportion problems. In the first method, the means (inner numbers) are multiplied together and the extremes (outer numbers) are multiplied together. If three values are known, the unknown value is found by using elementary algebra.

Example: $3:2 = 9:Z$ $\qquad$ $3Z = 18$ (2×9)

Means

Extremes $\qquad$ $Z = 6$

The second method of solving for the unknown value uses cross multiplication. In this method, the two ratios are expressed as fractions separated by an equal sign.

Example: $4:3 = 12:Z$ $\qquad \dfrac{4}{3} = \dfrac{12}{Z}$

The top portion of one ratio is multiplied by the bottom portion of the other ratio.

$\dfrac{4}{3} \qquad \dfrac{12}{Z}$ $\qquad$ $4Z = 36$ (3×12)

$Z = 9$

$3 \; \boxed{\text{X}} \; 12 \; \boxed{=} \; 36 \; \boxed{\div} \; 4 \; \boxed{=} \; 9$

PRACTICAL PROBLEMS

For problems 1–4 solve each equation for the value of the unknown.

1. $2:3 = 8\,\text{cm}:x\,\text{cm}$ $\qquad\qquad$ _____

2. $A:5 = 36\,\text{in.}:20\,\text{in.}$ $\qquad\qquad$ _____

3. $4:Y = 8\,\text{ft.}:40.5\,\text{ft.}$ $\qquad\qquad$ _____

4. $\dfrac{63}{28} = \dfrac{324\,\text{yd.}}{B}$ $\qquad\qquad$ _____

5. A pinion gear with 15 teeth turns at 200 rpm. It is driven by a larger gear
    having 60 teeth. Find the rpm of the larger gear.   _____

Note: Use these diagrams for questions 6–8.

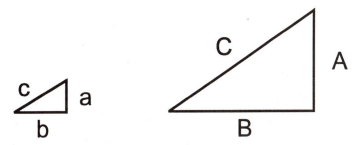

6. The sides of these two similar triangles are proportional. Side B = 27
    mm, side b = 9 mm, side a = 6 mm. What is the length of side A?   _____

7. If side a = 4 inches, side c = 9 inches, and side C = 42.75 inches, find
    side A.   _____

8. In the larger triangle, side B is 145 centimeters and side C is 180
    centimeters. In the smaller triangle, side b is 29 centimeters. Find, in
    centimeters, side c.   _____

9. A CAD operator realizes it takes 8 minutes to plot a drawing. How long
    will it take to plot 17 hard copies of the same drawing?   _____

10. A pump discharging 6 gallons of water per minute fills a tank in 30
    hours. How long does it take a pump discharging 20 gallons per minute
    to fill it?   _____

11. In two weeks, 5 men assemble 12 machines. How many men are
    needed to assemble 60 machines in the same time?   _____

12. A gear with 12 teeth and turning at 1275 rpm is driving a gear with 51
    teeth. Find the rpm of the larger gear.   _____

13. A gear with 91 teeth and running at 240 rpm is being driven by a gear
    turning at 840 rpm. How many teeth does the driver gear have?   _____

14. Two gears have a gear ratio of 2.8:1. If the larger gear has 84 teeth, how
    many teeth does the smaller gear have?   _____

15. The diameter of a driven pulley is 7.5 inches. The pulley is rotating at 180 rpm. The driver pulley is rotating at 375 rpm. Determine the diameter of the driver pulley.

_____

16. Pulley *A* is 9 inches, pulley *B* is 4 inches, pulley *C* is 11.25 inches. The rpm of pulley *A* is 352 and the rpm of pulley *D* is 2376. Find the diameter of pulley *D*.

_____

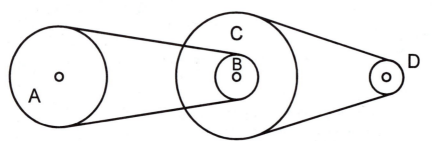

17. Find the missing term in the following proportion:
    $2\frac{1}{8} : 19\frac{1}{8} = ? : 38\frac{1}{4}$.

_____

18. A CAD operator realizes that it takes 40 minutes to plot a set of 6 drawings. Using the same set of 6 drawings:

    a. How many sets of drawings can be plotted in 480 minutes?

    _____

    b. How many drawings can be plotted in 480 minutes?

    _____

19. A structural drafter observes that a metal joint 8 feet long required 40 rivets. How many rivets are required for a joint 5 feet long?

_____

20. If 8 CAD drafters can produce an average of 24 drawings in 4 days, how many CAD drafters will it take to average 36 drawings in 4 days? Assume the same set of drawings will be produced.

_____

Applied Algebra

Unit 30 SYMBOLS AND EQUATIONS

BASIC PRINCIPLES OF SYMBOLS AND EQUATIONS

Symbols are a form of worldwide communication. They are used extensively on working drawings as a form of shorthand and a simplified way of denoting operations that need to be performed. They also are used to identify quantities and units of measurement. Symbols make it possible to develop mathematical formulas.

Quantities may be expressed in terms of numbers or a combination of numbers and symbols. Where letters are used in combination with numbers like $4A$ or $3B$, the expression is called a literal term.

Example: The perimeter of the rectangle is expressed in literal terms.

$$P = 2L + 2H$$

(rectangle with height labeled H and length labeled L)

Example: Express the sum of this example in terms of the literal factor.

$1B + 3B + (7B - 4B) + (5B - 3B) =$
$1B + 3B + 3B + 2B = 9B$
Answer: $9B$

Equations are used like symbols to solve problems that would otherwise be difficult and complicated. An equation is a statement in mathematical terms to indicate that the quantities or expressions on both sides of an equal sign (=) are equal. It also could be said that an equation is an equality of two quantities. Equations use a combination of numerical quantities and symbols to solve problems.

Example: What number added to 9 will equal 17?

$X + 9 = 17$
$8 + 9 = 17$
$X = 8$

The expressions appearing on both sides of the equal sign (=) are called members. Usually the quantity to the left of the sign is called the first member while the quantity to the right of the sign is called the second member.

The first member of the illustration is (3B + 4); the second member is (10).

$$3B + 4 = 10$$

PRACTICAL PROBLEMS

Using the given value for the variable, find the numerical value of each expression.

1. $N - 5$ when $N = 9$ _____

2. $5D$ when $D = 4 \frac{9}{16}$ _____

3. $\frac{s}{2}$ when $s = 7.54$ _____

4. $3x + 2y$ when $x = 4$; $y = 11 \frac{1}{4}$ _____

5. $3(a + b) + 2$ when $a = 6$; $b = 5$ _____

6. $A - \frac{B}{2} + 6$ when $A = 26.32$; $B = 8.12$ _____

Solve each equation for the value of the unknown.

7. $x + 9 = 13$ _____

8. $26 = 11.6 + y$ _____

9. $8Z - 4Z = 22$ _____

10. In terms of **T**, express dimension **A** on this strap.   _____

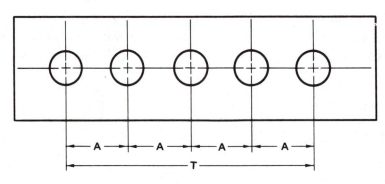

11. Express, in terms of **D,** the length of this template (dimension **L**).   _____

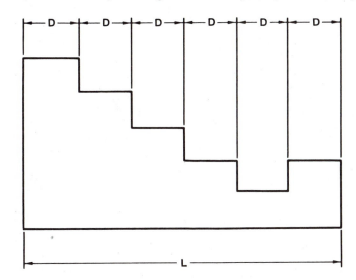

12. The length of the shortest side of a triangle is 7x. The longest side is 3.7 times the length of the shortest side. What is the length of the longest side in terms of x?   _____

13. Express, in terms of **L,** the literal dimensions **A** and **B** on this plate.   A _____

    B _____

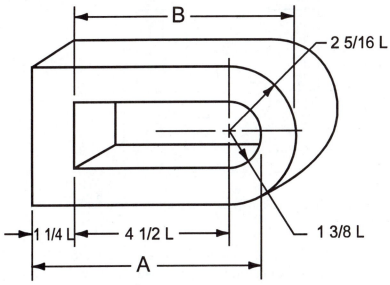

14. The side of a regular hexagon is 8H in length. What is the perimeter in terms of H?   _____

15. Find, in terms of x, the literal dimensions **A, B,** and **C** in this illustration.

A _____

B _____

C _____

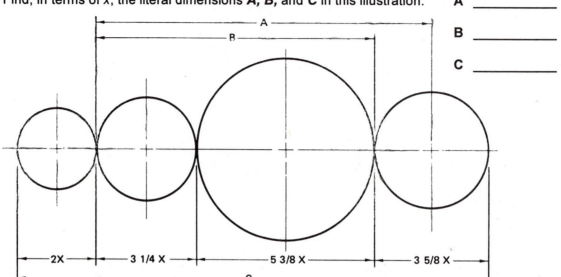

16. Find, in inches, the numerical value of x in this drawing. Determine dimensions **A, B, C,** and **D** of this bolt and nut.

X _____

A _____

B _____

C _____

D _____

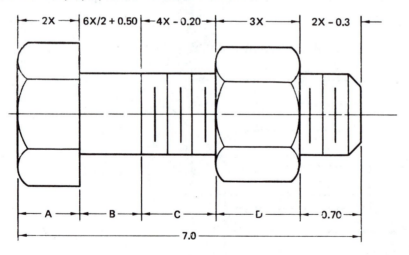

17. The circumference of a cylinder is 23.56 inches. What is the diameter to the nearest thousandth of an inch?

_____

18. A certain screw has 15 threads. This is $\frac{1}{6}$ as many threads as there are on a second screw. Find S, the number of threads on the second screw.

_____

19. A rectangular box is $24\frac{3}{8}$ inches wide. The length (L) is $17\frac{3}{16}$ inches longer than the width. What is the value of L in inches?

_____

20. In this drawing **A** = 1.50". Solve for the value of dimension **X**. _____

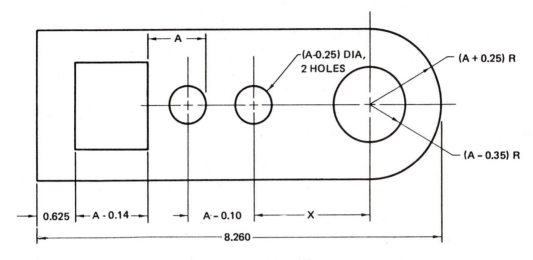

21. In this illustration, **B** = 1.25. Solve for the value of **Z**. _____

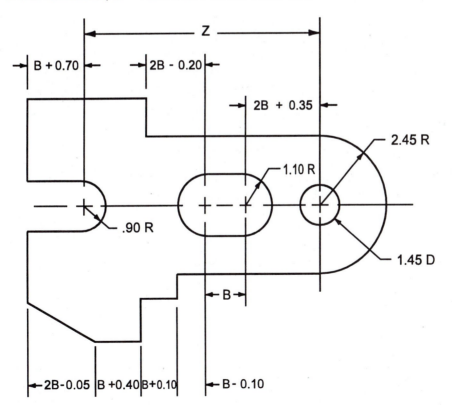

Unit 31 POWERS AND ROOTS

BASIC PRINCIPLES OF POWERS AND ROOTS

Powers and roots are often used in finding distances between different parts of machine drawings.

Powers or exponents is a convenient way to indicate the number of times a quantity is to be multiplied by itself. The exponent is the small number written to the right and slightly above the given quantity as shown.

$$4^2 = (4 \times 4) = 16 \qquad 3^3 = (3 \times 3 \times 3) = 27$$

4 (yˣ) 2 (=) 16   3 (yˣ) 3 (=) 27

A number can be raised to any power by using it as a factor that number of times as shown.

$$2^6 = (2 \times 2 \times 2 \times 2 \times 2 \times 2) = 64$$

2 (yˣ) 6 (=) 64

The root of a number is opposite its power. Determining the square root of a number means finding what number multiplied by itself will equal that number. The square root of 49 is 7. The square root of 49 is 7 because 7 squared equals 49. In finding the square root of a number, the number is placed under the radical sign ($\sqrt{\ }$) and no number is used outside the radical sign.

$$\sqrt{36} = 6$$

49 (√x) 7   36 (√x) 6

However, if a root other than a square root is to be determined, a small number is placed outside the radical sign to indicate what root is to be found.

$$\sqrt[3]{216} = 6$$

 216 (2ND) (ˣ√x) 6

PRACTICAL PROBLEMS

Raise each number to the given power.

1. $4^2 =$ _____

2. $8^2 =$ _____

3. $9^2 =$ _____

4. $6^3 =$ _____

5. $7^3 =$ _____

6. $5^4 =$ _____

7. $13^3 =$ _____

8. $5.36^2 =$ _____

9. $2.5^3 =$ _____

10. $1.5^4 =$ _____

11. $0.50^3 =$ _____

12. $(\frac{3}{4})^3 =$ _____

Find the square root of each number. Round answers to the nearest hundredth.

13. $\sqrt{49} =$ _____

14. $\sqrt{169} =$ _____

15. $\sqrt{4.375} =$ _____

16. $\sqrt{875} =$ _____

17. $\sqrt{1,898} =$ _____

18. $\sqrt{625} =$ _____

19. $\sqrt{0.8\%_{0.49}} =$ _____

20. $\sqrt{76.43} =$ _____

21. $\sqrt{217.5625} =$ _____

22. The area of a square is 42.09 square inches. Find the length of one side to the nearest hundredth of an inch. **Note:** *s* means side; $s = \sqrt{\text{Area}}$ _____

23. Find, in square inches, the area of this circle. Use the formula Area = $D^2 \times$ 0.7854, where *D* means diameter. (Round the answer to the nearest thousandth.)   _____

— 1.5 DIA

24. Find, in cubic millimeters, the volume of this cube. Use the formula $V = L^3$, where *L* means the length of the side. (Round the answer to the nearest thousandth.)   _____

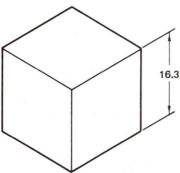

16.3

25. What is the length, in millimeters, of dimension C on this right triangle? Use the formula $C = \sqrt{A^2 + B^2}$ (Round the answer to the nearest thousandth.)

_____

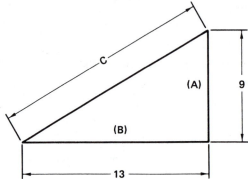

26. Find, in inches, the thickness (dimension A) of this piece of steel. Use the formula $A = \sqrt{B^2 - C^2}$ (Round the answer to the nearest thousandth.)

_____

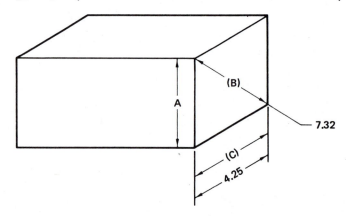

27. The left profile view of a step-up block is shown. What is the length of side B in inches? Use the formula $B = \sqrt{C^2 - A^2}$. (Round the answer to the nearest thousandth.)

_____

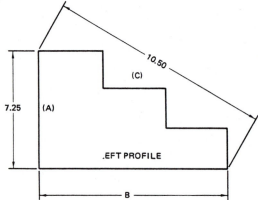

28. An architectural drafter needs to find the length of each side of a square room. The floor area of the room is 97.42 square meters. Find, in meters, the length of each side. Express the answer to the nearest hundredth.

_____

29. A mechanical drafter must find the actual length of line *A* on the part shown. Determine the distance *A*.

_____

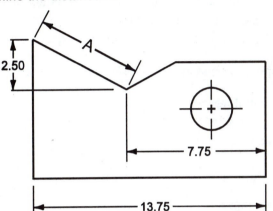

30. A landscape architect plans a circular reflection pool for a park. The area regulations will not permit a pool any larger than 254.4696 square meters. What is the largest permissible diameter that may be used for this reflection pool?

_____

Unit 32 FORMULAS AND HANDBOOK DATA

BASIC PRINCIPLES OF FORMULAS AND HANDBOOK DATA

Drafters often use formulas and handbook data while making drawings. Formulas are used to obtain dimensions for layout purposes or for the manufacture of a product and the construction of a building. Handbooks and technical publications contain technical information that must be referenced to determine dimensions, sizes, costs, weights, and so forth.

A formula is a mathematical statement of equality and is written using symbols.

Example: $C = \pi D$ C = Circumference
π = 3 ⅐ or 3.1416
D = Diameter

Examples:

1. Find the value of the sine of 38 degrees.
    Using the calculator:

    📟   38 (sin) .61566

2. Find the angle if the sine equals .61566.
    Using the calculator:

    📟   .61566 (2ND) (sin⁻¹) 37.99999 or 38°.

PRACTICAL PROBLEMS

1. Find the value of the sine of 78 degrees.   _____

2. Find the value of the tangent of 17 degrees.   _____

3. Find the value of the cosine of 59 degrees.   _____

4. Find the angle if the sine equals 0.74314.   _____

5. Find the angle if the tangent equals 9.5144.   _____

6. Find the angle if the cosine equals 0.92050.   _____

7. A drafter must determine several measurements before a gear can be drawn. These measurements are often found by using formulas. Most gear-cutting data can be found if the diametral pitch **P** and the number of teeth **N** are known. In this gear, the diametral pitch **P** is 4, and the number of teeth **N** is 24.

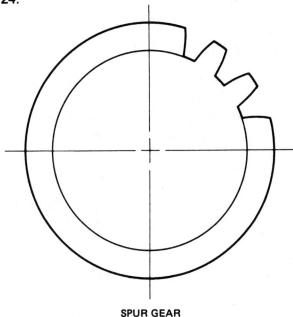

SPUR GEAR

a. Find, in inches, the pitch diameter **D**.

$$D = \frac{N}{P}$$

_____

b. Find, in inches, the outside diameter **O**.

$$O = \frac{N+2}{P}$$

_____

c. Find, in inches, the addendum **A**.

$$A = \frac{1.000}{P}$$

_____

d. Find, in inches, the thickness of the gear tooth **T**.

$$T = \frac{1.5708}{P}$$ (Round the answer to the nearest thousandth.)

_____

e. Find, in inches, the whole depth of the gear tooth **W**.

$$W = \frac{2.157}{P}$$ (Round the answer to the nearest thousandth.)

_____

8. What is the circumference of an 18-inch diameter circle?

    _____

9. What is the diameter of a circle with a circumference of 22.61 inches?

    _____

10. What is the diameter of a circle with an area of 18.1 square centimeters?

    _____

11. What is the circumference of a 7-millimeter diameter circle?

    _____

12. What is the radius of a circle with an area of 5,281 mm^2?

    _____

13. Drafters follow standard rules and a definite order in almost all work. When exact measurements are not known, approximate sizes and formulas are used. This $\frac{3}{4}$" diameter hexagonal bolt and nut are drawn from the approximate formulas given. Sizes are based on the diameter **D** of the fastener. State your answers in decimal form.

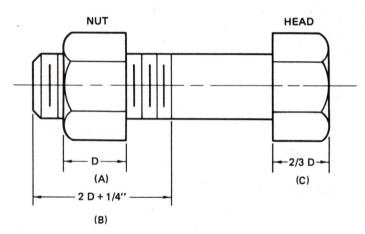

a. What is the threaded length of the bolt **B**?

    _____

b. What is the thickness of the nut **A**?

    _____

c. What is the thickness of the bolt head **C**?

    _____

Use TABLE I for equivalent units of measure.

14. What is the millimeter equivalent of 5 inches?

    _____

15. What is the inch equivalent of 8 millimeters?

    _____

16. What is the millimeter equivalent of 9 inches?

    _____

17. What is the area equivalent, in square centimeters, of 6 square inches?

    _____

18. What is the area equivalent, in square inches, of 4 square centimeters? _____

19. The formula for the minimum depth when tapping a thread in cast iron is 1 $\frac{1}{2}$ **D**. What is the minimum depth for tapping a $\frac{5}{8}$ - 16 NC thread in cast iron? _____

20. Find, in inches, the circumference of a 4-inch diameter disk cam. The circumference of the cam equals the stretchout length of this displacement diagram. (Round the answer to the nearest thousandth.) _____

DISPLACEMENT DIAGRAM

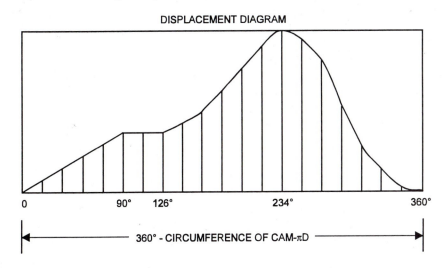

21. What is the square root of 23? _____

22. What is the square of 14 cm? _____

23. What is the cube root of 20 cubic inches? _____

24. What is the cube of 56 mm? _____

25. Find, in millimeters, the diagonal of this square. Diagonal = (1.414) (*s*); where *s* = length of side. (Round the answer to the nearest thousandth.) _____

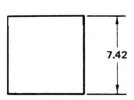

26. Find, in inches, the diameter **D** of a circle when the area **A** is 628.32 sq.
in. $D = \sqrt{\dfrac{A}{0.7854}}$

_____

27. When making a full development of a right circular cone the drafter lays
out an angle found by using the formula: Slope = $\frac{R}{S}$ × 360°. Determine the angle when **R** = 100 mm and **S** = 180 mm.

_____

28. Find, in inches, the amount of taper per inch **T** for this tapered shaft.
$T = \dfrac{D - d}{L}$ (Round the answer to four decimal places.)

_____

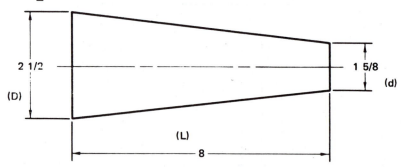

2 1/2 (D)   1 5/8 (d)   (L)   8

29. Regular polygons have equal angles and sides of equal length. Find the
number of degrees in angles **A, C,** and **E**.

A _____

C _____

E _____

N = Number of sides

$\angle C = \dfrac{360°}{N}$

$\angle A = \dfrac{N - 2}{N} \times 180$

$\angle E = \angle C$

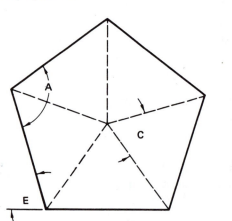

30. Determine the area of a stretchout (development) of a right octagonal
prism having a side dimension of 1.17 inches and a height of 4.45
inches. Use the formula $A = 4S \times H$.

_____

Graphs

Unit 33 USE OF GRAPHS

BASIC PRINCIPLES OF GRAPHS

A graph is a diagram that allows the comparison of data presented on two or more variables at the same time. The most commonly used graphs include bar graphs, line graphs, and circle graphs (pie charts). The type of graph used to compare data would depend on the nature of the data to be presented.

PRACTICAL PROBLEMS

1. The total expenses of a company are represented in the following circle graph.

Company Expenditures

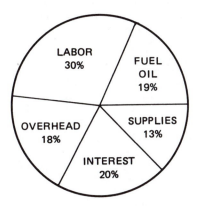

a. Which part represents the largest expense of the company?   _____

b. Which part represents the smallest expense of the company?   _____

c. Find the percent which represents the difference between the largest and smallest expenses.   _____

2. The line graph illustrates the appliance shipments made during a 12-month period.

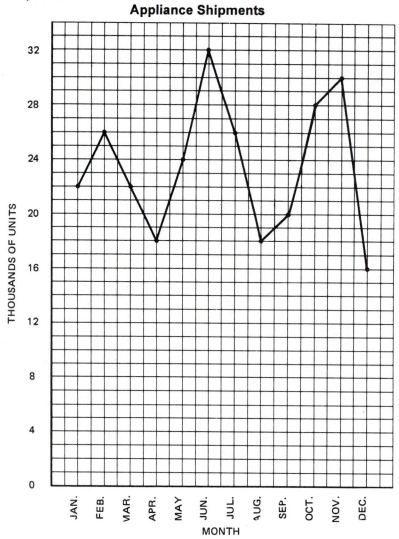

Appliance Shipments

a. Over what periods of time did the shipments decrease the most? _____

b. In which month did the greatest amount of shipments take place? _____

c. In which month did the least amount of shipments take place? _____

d. In which month was the shipment halfway between the lowest and highest shipments? _____

e. During which month were 20,000 units shipped? _____

3. The production for the six-year period is shown on this bar graph.

Six-Year Production

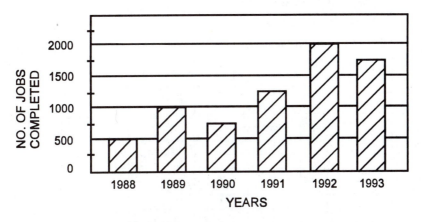

a. In which year were the greatest number of jobs completed? _____

b. Which year represents the smallest production? _____

c. In which year was the number of jobs completed halfway between the highest production and lowest production years? _____

d. What was the difference in production for years 1988 and 1993? _____

4. The United States raw steel production by types is: open hearth—57%; electric furnace—12%; basic oxygen—31%. Using the circle given, develop a circle graph illustrating this information.

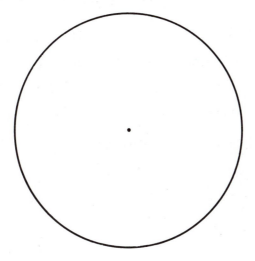

5. The graph shown provides the number of hours per month that are spent on a special company drafting project.

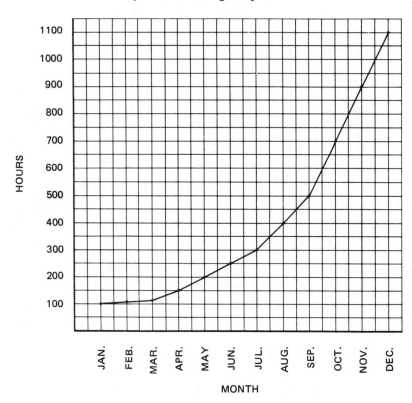

**Annual Report of Hours Spent
on Special Drafting Project**

a. How many hours are spent during the month of November? _____

b. In what month are the hours spent seven times the number of hours spent in January? _____

c. In what month are 150 hours spent on the project? _____

d. Which three months have the largest increase in hours? _____

_____

_____

6. Develop a bar graph to show the personnel make-up of the following engineering department.

drafters	24%	engineers	14%
checkers	10%	secretarial help	6%
technicians	12%	clerks	6%
architects	10%	CAD operators	18%

Personnel in Engineering Department

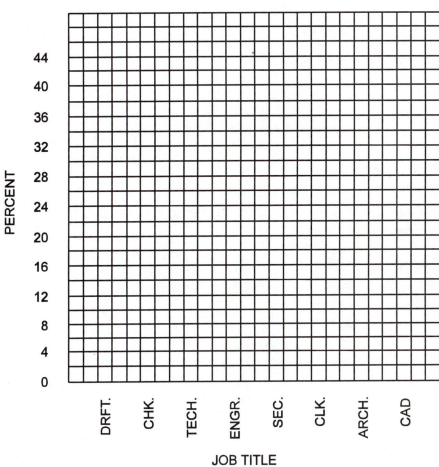

7. Develop a line graph to show a summer audit of a company's production. In April the net gain was $15,000; May—$20,000; June—$10,000; July—$25,000; August—$30,000; and September—$27,500.

Net Gain in Production at Acme Wholesale Company

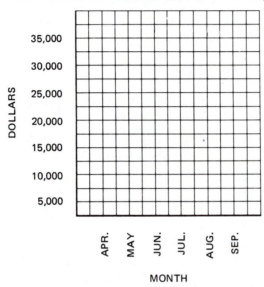

8. Using the circle given, develop a circle graph illustrating the type of drawings done by a drafting department.

Machine drawings	37%	Foundry drawings	23%
Structural drawings	18%	Sheetmetal drawings	15%
Pattern drawings	7%		

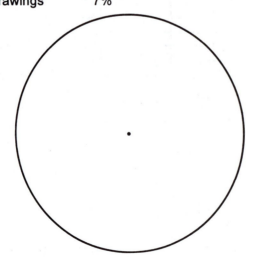

Applied Geometry

Unit 34 LINES, SHAPES, AND GEOMETRIC CONSTRUCTION

BASIC PRINCIPLES OF LINES, SHAPES, AND GEOMETRIC CONSTRUCTION

Almost all objects can be broken down into geometric shapes. In order to make correct drawings, the drafter must be able to construct different types of polygons, pyramids, and prisms.

A knowledge of basic geometric constructions using a compass and an appropriate scale is required by all drafters as they lay out their work. CAD operators will find that some geometric constructions are accomplished in a different manner depending on the computer software being used. In any event, one must visualize the operation to be performed.

PRACTICAL PROBLEMS

1. What type of angle is ∠ **AOB**?   _____

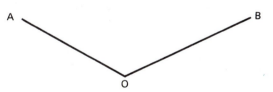

2. What type of angle is ∠ **COD**?   _____

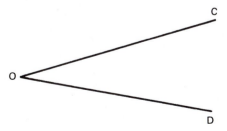

3. How many degrees are there in a right angle?   _____

4. Which two of these triangles are congruent?   _____

_____

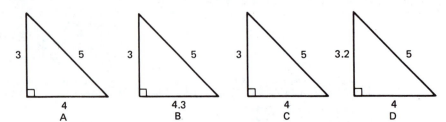

5. Identify each geometric figure using names, such as "isosceles triangle."   A _____

B _____

C _____

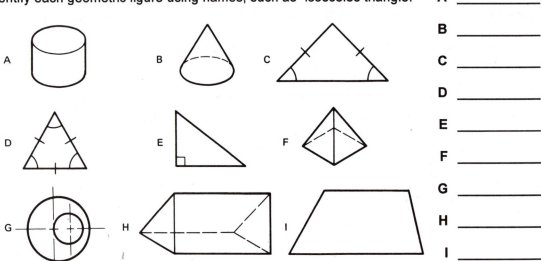

D _____

E _____

F _____

G _____

H _____

I _____

Note: In the following problems, show all construction lines.

6. Bisect line **AB** using a compass.

7. Bisect arc **CD** using a compass. **Hint:** Draw segment **CD**. The bisector of $\overline{CD}$ also bisects $\overset{\frown}{CD}$.

8. Bisect angle **AOB** using a compass.

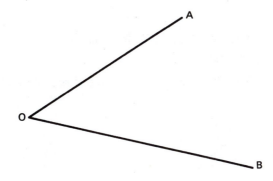

9. Divide line **CD** into seven equal parts.

C ——————————————————— D

10.   Using a compass, drop a perpendicular from point **L** to line **MN**.

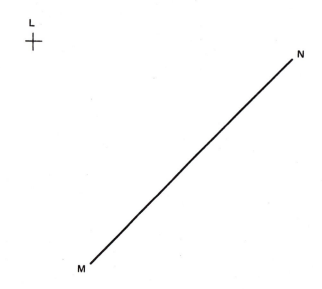

11.   a.   Construct a tangent to the circle at point **T**. Indicate the one-inch segment which has **T** as its center.

b.   Construct a tangent from **P** to the lower part of the circle and label the point of tangency.

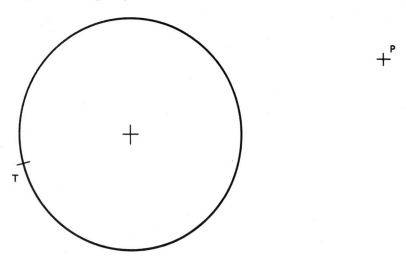

12. Inscribe a regular hexagon in an 84-mm diameter circle.

13. Using line **AB** as the diagonal, construct a square.

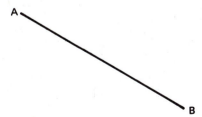

14. Construct a regular hexagon using the line **CD** as "across the flats" distance. The distance *across the flats* is the measurement across any set of parallel sides.

C ───────────── D

15. Construct an equilateral triangle with line **EF** as the base.

E ───────────── F

16. Construct a regular octagon with line **GH** as the "across the corners" distance. The distance *across the corners* is the measurement from the vertex of one pair of sides to the vertex of the opposite pair of sides.

G ──────────────── H

17. Construct a circle through points **A, B,** and **C**.

Step 1. Connect **A** and **B**; connect **B** and **C**.
Step 2. Bisect $\overline{AB}$; bisect $\overline{BC}$.
Step 3. With compass point at intersection of bisectors, and distance to **A** as a radius, draw the circle.

A +

+ C

B +

18.   Draw an arc tangent to line **AB**, at **B**, and passing through point **C**.

Step 1.   Draw $\overline{BC}$.
Step 2.   Bisect $\overline{BC}$.
Step 3.   Construct a perpendicular at **B**.
Step 4.   Use the intersection of these lines as the center of the circle.
Use the distance to **B** as the radius. Draw the arc.

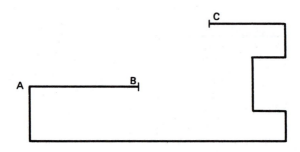

19.   Construct a square using line **CD** as the base.

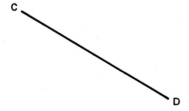

20. Construct triangle **ABC** using line **AB** as the base. Make side
    AC = 45 mm and side **BC** = 89 mm.

    Step 1. Use 45 mm as a radius and draw an arc.
    Step 2. Use 89 mm as a radius and draw an arc.
    Step 3. The intersection of these arcs is **C**.

Note: The following problems are more difficult. They require the application of many basic constructions.

21. Draw a 25.40 mm diameter hole equidistant from centers **A** and **B** and
    equidistant from centers **C** and **D**.

    Step 1. Connect **A** and **B**, then construct its perpendicular bisector.
    Step 2. Connect **C** and **D**, then construct its perpendicular bisector.
    Step 3. Use the intersection of these lines as the center of the circle.

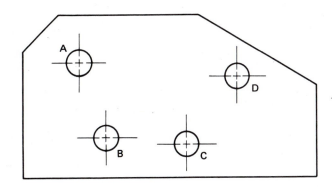

22. Use full scale to lay out this plate. (Do not use a protractor.)

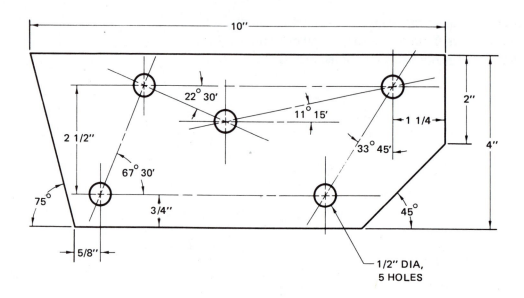

23. Lay out the sets of holes following the given instructions.

    a. Bisect ∠**B** and construct three equally spaced ¼" diameter holes. Make the first hole ¾" from point **B** and the last hole 2⁵⁄₁₆" from point **B**.

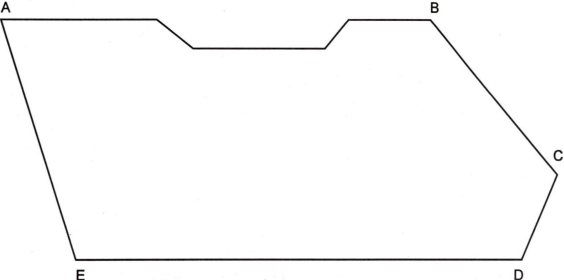

    b. Bisect ∠**A** and construct four equally spaced ⅜" diameter holes. Make the first hole ⅞" from point **A** and the last hole 3⁹⁄₁₆" from point **A**.

    c. Construct a ¾" diameter circle 1³⁄₁₆" from point **C** and 1⅜" from point **D**.

24. Construct an arc of 1 ¼-inch radius tangent to the given arc and to the straight line. Show all construction marks. Label the tangent points.

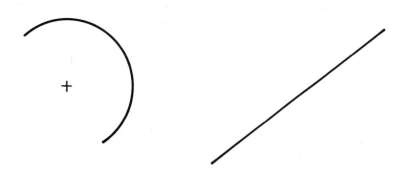

25. Construct an arc of 42-mm radius tangent to the two given arcs. Show all construction marks. Label the tangent points.

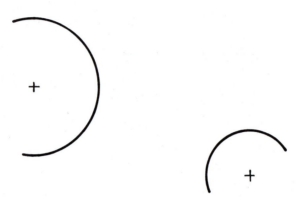

26. Construct an arc of one-inch radius tangent to both the inside of arc **A** and the outside of arc **B**. Show all construction lines and mark the tangent points.

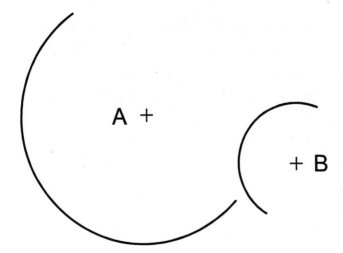

27. Using the partial view given, construct an instrument drawing identical to the example provided. Use the sizes indicated in the example and mark the tangent points where directed by your instructor.

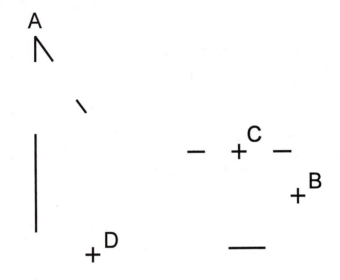

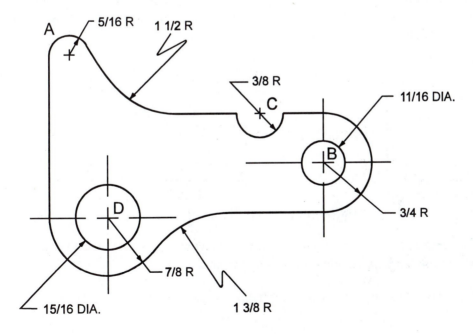

EXAMPLE

Applied Trigonometry

Unit 35 RIGHT TRIANGLES

BASIC PRINCIPLES OF RIGHT TRIANGLES

Trigonometry is a branch of mathematics that deals with finding the unknown values of the sides and angles of triangles. When dealing with right triangles, drafters and CAD operators must understand the mathematical principles and trigonometry operations underlying the problem. Trigonometry is very important to the work of all drafters.

A right triangle, as the name implies, is a triangle containing a right angle.

Description of Sides

The sides of a right triangle are named opposite side, adjacent side, and hypotenuse. The hypotenuse is always the side opposite the right angle. The positions of the opposite and adjacent sides depend on the reference angle used. The opposite side is opposite the reference angle and the adjacent side is next to the reference angle. Identification of these sides is important regardless of the position of the triangle itself.

Example:

HYPOTENUSE SIDE OPPOSITE A SIDE ADJACENT

HYPOTENUSE B SIDE ADJACENT SIDE OPPOSITE

Trigonometric Functions

There are six terms that are used to express the ratios between the sides of a triangle. These ratios are the sine, cosine, tangent, cotangent, secant, and cosecant.

The sine (sin) is the ratio of the side opposite to the hypotenuse. In triangle ABC, the

$$\sin A = \frac{\text{side opposite}}{\text{hypotenuse}} = \frac{a}{c}$$

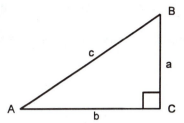

Using the same triangle, the cosine (cos) is the ratio of the side adjacent to the hypotenuse.

$$\cos A = \frac{\text{side adjacent}}{\text{hypotenuse}} = \frac{b}{c}$$

In the same example, the tangent (tan) is the ratio of the side opposite to the side adjacent.

$$\tan A = \frac{\text{side opposite}}{\text{side adjacent}} = \frac{a}{b}$$

The cotangent (cot) is the ratio of the side opposite to the side adjacent.

$$\cot A = \frac{\text{side adjacent}}{\text{side opposite}} = \frac{b}{a}$$

The secant (sec) and cosecant (csc) of angle A may be expressed in algebraic form as well.

$$\sec A = \frac{\text{hypotenuse}}{\text{side adjacent}} = \frac{c}{b} \qquad \csc A = \frac{\text{hypotenuse}}{\text{side opposite}} = \frac{c}{a}$$

Functions and Cofunctions

In the triangle ABC, angle A and B are complimentary and their sum equals 90 degrees. The term *cofunction* refers to the function of the complimentary angle.

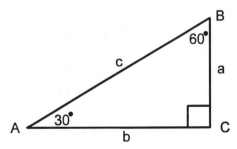

The cofunction of the sine is the cosine, the cofunction of the tangent is the cotangent, and the cofunction of the secant is the cosecant.

In the triangle ABC, the sine of 30 degrees (.50000) equals the cosine of 60 degrees (.50000).

Example: Determine the angles in the right triangle ABC whose sides equal 10", 8", and 6".

sin C = $\dfrac{8}{10}$ = .80000 = 53 degrees

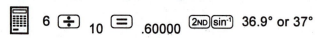 8 ⊕ 10 ⊜ .80000 (2ND)(sin⁻¹) 53°

sin A = $\dfrac{6}{10}$ = .60000 = 37 degrees

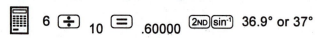 6 ⊕ 10 ⊜ .60000 (2ND)(sin⁻¹) 36.9° or 37°

Complimentary angles 53 degrees and 37 degrees total 90 degrees

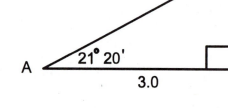

Example: Determine the side X of the right triangle ABC.

tan A = $\dfrac{\text{side opposite}}{\text{side adjacent}}$

tan 21 degrees 20' = $\dfrac{X}{3.0}$

(From table) .39055 = $\dfrac{X}{3.0}$

.39055 × 3.0 = X
   1.17 = X

▦ (21° 20') = 21.33333 (tan) .39055 (X) 3.0 ⊜ 1.17

Pythagorean Theorem

The pythagorean theorem can be used to find the length of a side of a right triangle when the lengths of the other two sides are known. The rule states that the square of the hypotenuse is equal to the sum of the squares of the remaining sides. The hypotenuse is the side opposite the right angle.

$$c^2 = a^2 + b^2$$

If the value of c is to be found, the formula is expressed as follows.

$$c = \sqrt{a^2 + b^2}$$

If the hypotenuse and one side are known, the following formulas can be used.

$$a = \sqrt{c^2 - b^2} \quad \text{or} \quad a^2 = c^2 - b^2$$
$$b = \sqrt{c^2 - a^2} \quad \text{or} \quad b^2 = c^2 - a^2$$

Example: Determine the length of side c in the right triangle given.

$$c^2 = a^2 + b^2$$
$$c^2 = 3^2 + 4^2$$
$$c^2 = 9 + 16$$
$$c^2 = \overline{25}$$
$$c = \sqrt{25}$$
$$c = 5$$

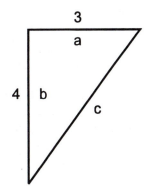

Example: Determine the length of side a in the right triangle presented.

$$a^2 = c^2 - b^2$$
$$a^2 = 12^2 - 7^2$$
$$a^2 = 144 - 49$$
$$a^2 = \overline{95}$$
$$a = \sqrt{95}$$
$$a = 9.75$$

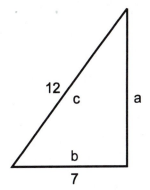

▦ 12 (X²) 144 (—) 7 (X²) 49 (=) 95 (√X) 9.75

PRACTICAL PROBLEMS

1. What is the cofunction of the sine of 37 degrees? _____

2. What is the cofunction of the cotangent of 21 degrees? _____

3. What is the cofunction of the cosine of 78 degrees? _____

5. What is the angle, in degrees and minutes, with a tangent value of 4.4494?

    _____

6. Find the sine of 27° 13'.

    _____

7. Find the cosine of 27° 13'.

    _____

8. Find, in degrees, angle **X**. All dimensions are in inches.

    _____

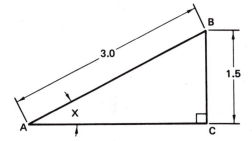

9. Find, in inches, the length of side **FG**.

    _____

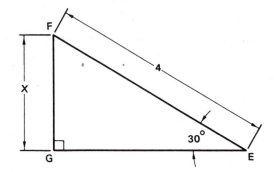

10. In this right triangle, side **BC** equals 10 inches and side **AB** equals 15 inches. Find the length of side **AC** to the nearest hundredth inch.

    _____

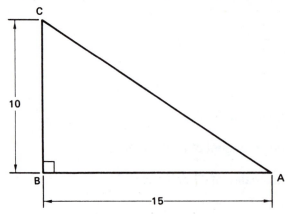

11. Find, to the nearest thousandth inch, dimension **X** of this shelf brace. _____

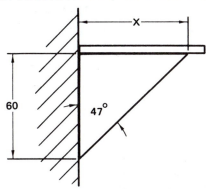

12. What is the base length of a right triangle with a 35° base angle and a 16-inch altitude (height)? Express answer to the nearest hundredth inch. _____

13. What is the hypotenuse of a right triangle with a 60° base angle and a 14-inch base length? _____

14. Find, to the nearest hundredth inch, the unknown side of this template. _____

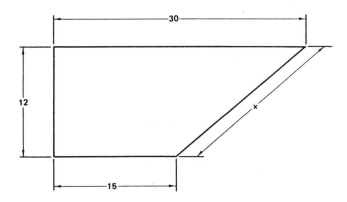

15. Find the height **B** of this gusset plate. _____

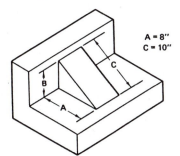

16. Find the center distance (between centers) from hole **A** to hole **C**; from hole **B** to hole **C**. Express answers to the nearest thousandth inch.

    AC _____

    BC _____

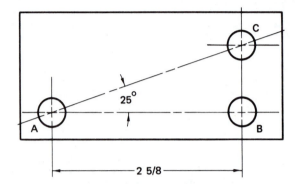

17. What angle (**X**) is needed to machine this shaft as illustrated? Express answer to the nearest minute. All dimensions are in inches.

    _____

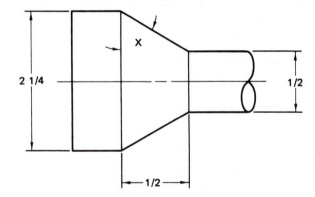

18. Find the center distance **X**, to the nearest hundredth inch, on this gasket.

    _____

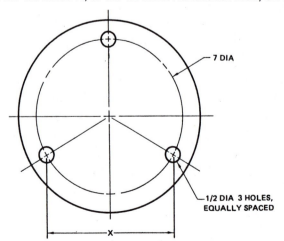

19. Find the missing dimension **X**, to the nearest hundredth millimeter, on this template.

_____

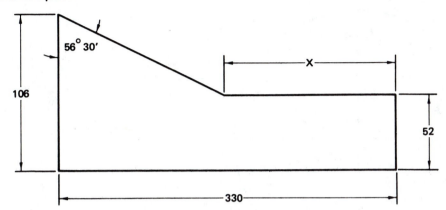

20. Determine the distance **X**, to the nearest hundredth inch, on this match plate.

_____

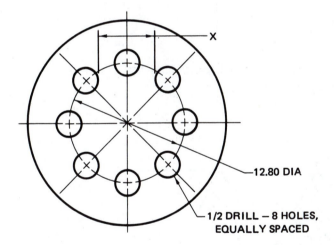

21. Find the distance **X**, to the nearest hundredth inch, on this circular plate. _____

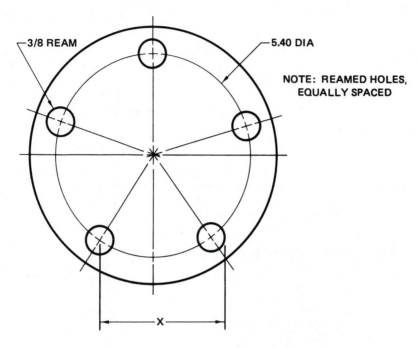

-3/8 REAM   -5.40 DIA

NOTE: REAMED HOLES,
     EQUALLY SPACED

X

22. Given is an American National thread form. Find the depth **D** of the thread to the nearest thousandth inch. _____

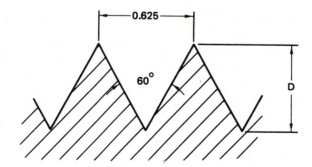

0.625

60°   D

23. Find distance **X**, to the nearest thousandth inch, on the gauge illustrated. _____

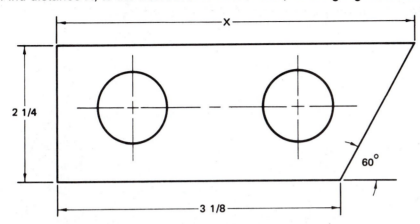

24. A drafter is designing a template. It must be rotated on pivot hole **A** to punch the other three holes. Determine dimension **X** to the nearest thousandth inch. _____

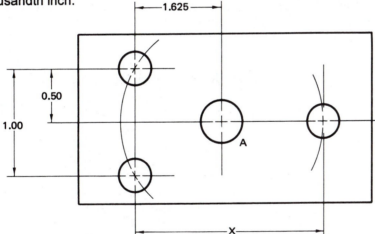

25. Find, to the nearest thousandth inch, the distance across the crests of the acme thread. _____

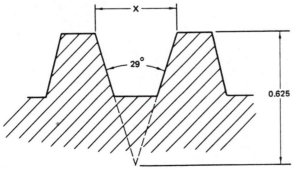

26. A drafter is designing a gauge to accurately measure the distance between pins. Find the missing dimensions **A** and **B** to the nearest thousandth inch.

A _____

B _____

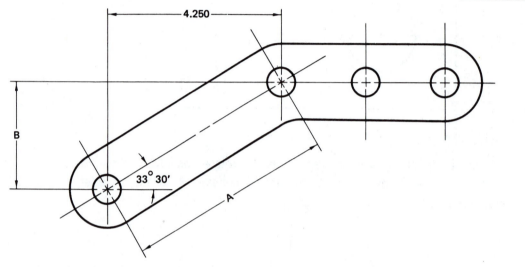

27. Determine the length **L**, to the nearest thousandth inch, of the taper plug illustrated.

_____

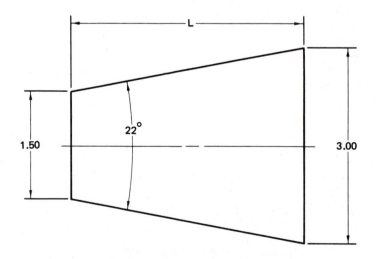

28. Determine the grid spacing for the CAD drawing presented.   _____

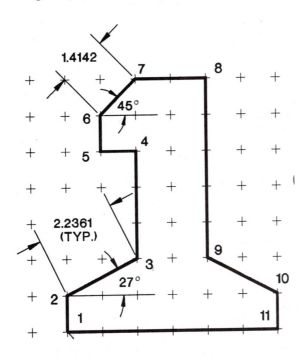

Reprinted with permission from Stellman, Krishnan, and Rhea,
Harnessing AutoCAD, copyright 1993 by Delmar Publishers.

29. Find the inclined distance between holes **A** and **B** using the CAD drawing shown. Express your answer to the nearest hundredth inch.   _____

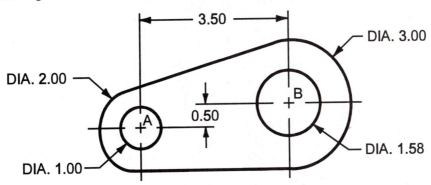

Reprinted with permission from Stellman, Krishnan, and Rhea,
Harnessing AutoCAD, copyright 1993 by Delmar Publishers.

30. Determine the vertical height of the auxiliary plane **A** on the CAD drawing illustrated. Express your answer to the nearest thousandth inch.   _____

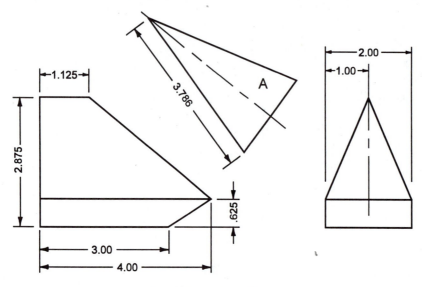

31. Using the CAD drawing presented, find the length of line **A**. If angle **B** is 30° 30', determine the length of line **C**. Express your answer to the nearest hundredth.

A   _____

B   _____

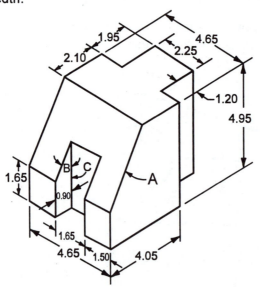

32. Determine the lengths of lines **A** and **B** to the nearest hundredth using the CAD drawing shown.

A _____

B _____

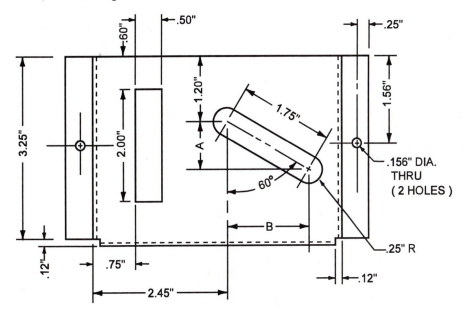

33. Find the length of line **A** to the nearest thousandth using the CAD drawing illustrated.

_____

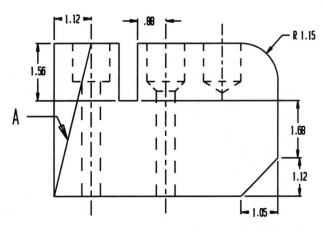

CAD DRAWING – LOCATING BLOCK

34. Determine the overall height and width to the nearest hundredth using the CAD drawing given.

Height _____

Width _____

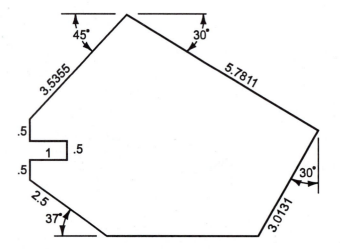

35. Determine the lengths of lines **A, B,** and **C** to the nearest hundredth using the CAD drawing presented.

A _____

B _____

C _____

Unit 36 OBLIQUE TRIANGLES

BASIC PRINCIPLES OF OBLIQUE TRIANGLES

Oblique triangles are as important to all drafters as right triangles. An oblique triangle does not contain a right angle, but a series of right triangles can be established. The use of the sine and cosine laws simplifies the computations when working with oblique triangles. Drafters should utilize these formulas when applicable.

The Law of Sines

The law of sines states that in any triangle, the sides are proportional to the sines of the opposite angles. The law of sines is used to solve the following types of problems:

- where only one side and any two angles are known;
- where any two sides and an angle opposite one of the given sides are known.

In triangle ABC, the law of sines would be expressed as follows:

$$\frac{a}{\sin A} = \frac{b}{\sin B} = \frac{c}{\sin C}$$

Example: Given two angles and a side, find side x in the oblique triangle DEF. You would start by setting a proportion to solve for x.

$$\frac{x}{\sin 41°} = \frac{3.25}{\sin 53°}$$

$$\frac{x}{.65606*} = \frac{3.25}{.79864*}$$

* Calculator is used to find sin values.

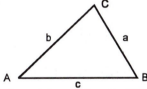

41 (sin) .65606
53 (sin) .79864

.79864x = 3.25 (.65606)

x = $\frac{3.25\,(.65606)}{.79864}$

x = 2.67 inches

  3.25 (X) .65606 (÷) .79864 (=) 2.67

The Law of Cosines

The law of cosines states that in any triangle, the square of any side is equal to the sum of the squares of the other sides minus twice the product of the two sides and the cosine of their included angle. The law of cosines is used to solve the following types of problems:

- when two sides and an included angle are known;
- when three sides are known.

In terms of triangle ABC, the following formulas are used:

$$a^2 = b^2 + c^2 - 2 \, bc \cos A$$

$$b^2 = a^2 + c^2 - 2 \, ac \cos B$$

$$c^2 = a^2 + b^2 - 2 \, ab \cos C$$

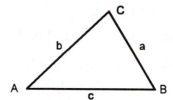

Transposing these equations the cosine values become:

$$\cos A = \frac{b^2 + c^2 - a^2}{2 \, bc} \qquad \cos B = \frac{a^2 + c^2 - b^2}{2 \, ac} \qquad \cos C = \frac{a^2 + b^2 - c^2}{2 \, ab}$$

Example: In triangle ABC, two sides and the included angle are given. Determine side x.

Use the formula $a^2 = b^2 + c^2 - 2 \, bc \cos A$

$a^2 = 4^2 + 9^2 - 2 \times 4 \times 9 \times .89879$

$a^2 = 16 + 81 - 8 \times 9 \times .89879$

$a^2 = 97 - 72 \times .89879$

$a^2 = 97 - 64.71$

$a = \sqrt{32.29}$

$a = 5.68$ inches

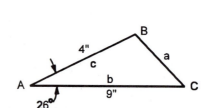

 26 (cos) .89879

4 (X²) 16 (+) 9 (X²) 81 (=) 97
4 (X) 9 (X) 2 (X) .89879 (=) 64.71
97 (−) 64.71 (=) 32.29 (√x) 5.68

PRACTICAL PROBLEMS

1. What is the sine of 116° 17'? _____

2. An oblique triangle contains angles of 17° 37' 23" and 109° 26' 47". Determine the third angle. _____

3. Determine the cofunction of the sine of 35° 27'. _____

4. The sum of two angles of a triangle equal 127° 43' 19". What is the value of the third angle? _____

5. Determine the tangent of 142° 13'. _____

6. Given a scalene triangle having angles of 35° 34' and 83° 47', determine the third angle. _____

7. What is the cosine of 147°? _____

8. Determine the tangent of 127° 41'. _____

9. Find, to the nearest hundredth inch, the length of side *a* and side *b* on this triangle.

    a _____

    b _____

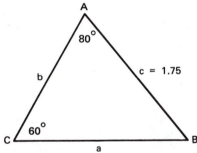

10. Find, to the nearest hundredth inch, the length of side *a*. Find, to the nearest ten seconds, angle **C** of the triangle.

    a _____

    C _____

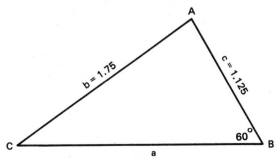

11. Find, to the nearest hundredth inch, side **X** of the triangle.   _____

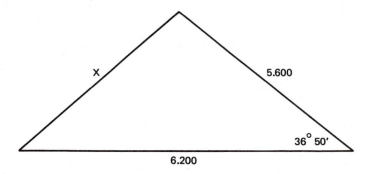

5.600

X

36° 50′

6.200

12. A mechanical drafter must give the length between points **B** and **C** on a detail drawing. The distance between points **A** and **C** is 1.1019 inches and between points **A** and **B** is 1.398 inches. The angle at **C** is 53°. What is the length between points **B** and **C**? Express answer to the nearest thousandth inch. **Hint:** A simple drawing of this problem is helpful.

_____

13. Find, to the nearest minute, the angles of this triangle. All dimensions are in inches.

A _____

B _____

C _____

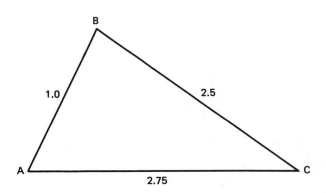

B

1.0

2.5

A

2.75

C

14. Find the distance between the centers of holes **B** and **C** using this circular plate illustrated. The distance between the centers of **A** and **C** is 1.3838 inches. Express answer to four decimal places. _____

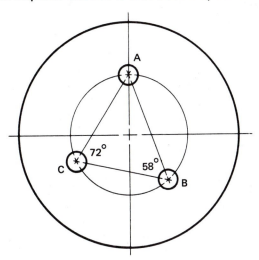

15. Find, to the nearest hundredth inch, dimension **X** on this template. _____

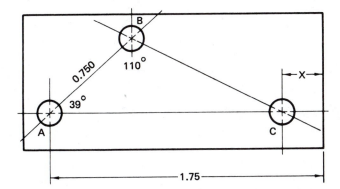

16. What is the distance between the centers of holes **C** and **A** on this layout fixture? Express answer to nearest hundredth inch. _____

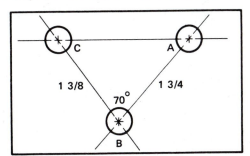

17. Find, to the nearest degree, angle **X** (the obtuse angle) in this triangle. All dimensions are in inches.

_____

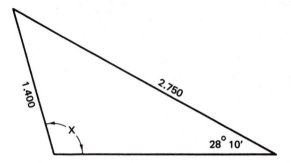

18. Find, to the nearest minute, angle **B** on this triangle. All dimensions are in inches.

_____

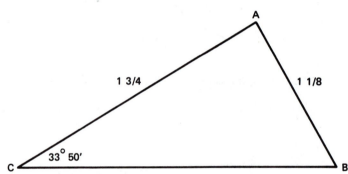

19. Find, to the nearest minute, angle **A** in this diagram. All dimensions are in inches.

_____

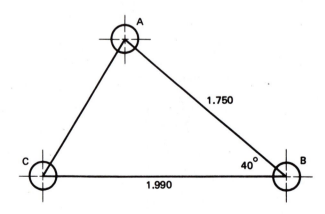

20. Find, to the nearest hundredth inch, the distance **X** on this plate. _____

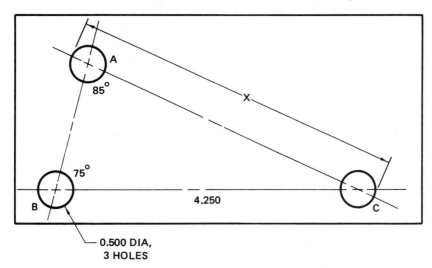

0.500 DIA,
3 HOLES

21. Determine the lengths of lines **A, B,** and **C** to the nearest thousandth using the CAD drawing presented.

A _____

B _____

C _____

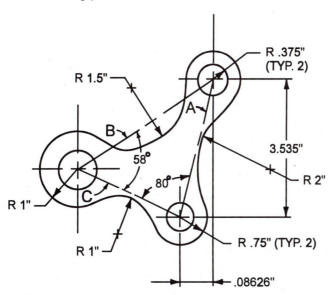

Reprinted with permission from Stellman, Krishnan, and Rhea,
Harnessing AutoCAD, copyright 1993 by Delmar Publishers.

22. Find the lengths of lines **B** and **C** if line **A** is 3.536 using the CAD drawing below. Express your answer to the nearest hundredth.

B _____

C _____

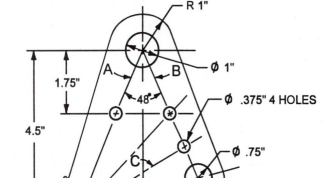

Reprinted with permission from Stellman, Krishnan, and Rhea,
Harnessing AutoCAD, copyright 1993 by Delmar Publishers.

23. Determine the length of line **AB** to the nearest hundredth using the CAD drawing presented. The angle at **C** equals 76°.

AB _____

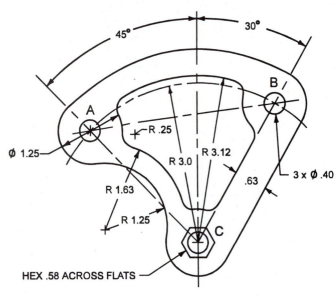

Reprinted with permission from Resetarits and Bertolini,
Using CADKEY Light, copyright 1992 by Delmar Publishers.

24. Find the lengths of lines **B** and **C** using the CAD drawing illustrated. Line **A** is 3.536 long. Express your answer to the nearest thousandth.

B _____

C _____

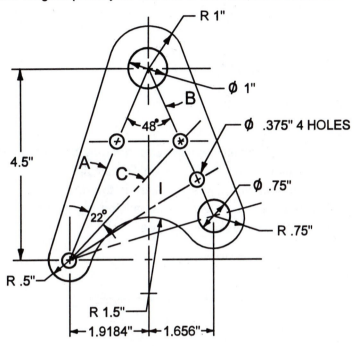

Appendix

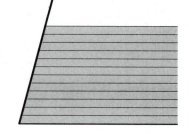

Section I

Using the Scientific Calculator

There are numerous brands and types of scientific calculators that may be purchased to take the drudgery out of mathematical calculations. The TI-30X scientific calculator is being explained in the following examples. Common functions and questions used in the related Drafting and CADD fields such as squaring a number, finding square roots, determining trigonometric functions, and so forth are illustrated using the following keystroke sequences.

KEY	FUNCTION
ON/C	Turns the calculator on, also used to clear the display.
OFF	Turns the calculator off.
2nd	Accesses the second function of the next key pressed (function above the key).
SCI	Sets display to scientific notation.
ENG	Sets display to engineering notation.
DRG	Sets angle unit to degrees, radians or gradient.
(	Opens parentheses.
)	Closes parentheses.
FIX	Fixes the number of significant digits.
STO	Stores the displayed value in memory.

RCL   Recalls the value stored in memory.

a b/c   Enters the proper or improper fraction b/c.

F ⇔ D   Toggles display number between fractional and decimal form.

d/c   Toggles displayed fraction between a mixed number and an improper fraction.

+   Performs addition.

−   Performs subtraction.

×   Performs multiplication.

÷   Performs division.

=   Closes any open parenthetical expression and completes all pending operations.

+/−   Changes the sign of the displayed number.

X !   Calculates the factorial of the displayed value.

π   Enters the value of pi, rounded to ten digits.

→   Removes the last digit from a number entry.

%   Calculates percentages, ratios, add-ons, and discounts.

x²   Squares the displayed value.

x³   Cubes the displayed value.

yˣ   Raises the displayed value to a specified power.

ˣ√   Finds a specified root of the displayed value.

(√X)	Calculates the square root of the displayed value.
(1/x)	Calculates the reciprocal of the displayed value.
(LOG)	Calculates the common logarithm of the displayed value.
(DMS ► DD)	Converts displayed value to equivalent decimal degrees.
(DD ► DMS)	Converts displayed value to a decimal-degree angle.
(SIN)	Calculates the sine of the displayed angle.
(COS)	Calculates the cosine of the displayed angle.
(TAN)	Calculates the tangent of the displayed angle.

Section II

EQUIVALENT TABLES

TABLE I
EQUIVALENT ENGLISH AND METRIC UNITS OF MEASURE

Linear Measure

Unit	Inches to millimetres	Millimetres to inches	Feet to metres	Metres to feet	Yards to metres	Metres to yards	Miles to kilometres	Kilometres to miles
1	25.40	0.03937	0.3048	3.281	0.9144	1.094	1.609	0.6214
2	50.80	0.07874	0.6096	6.562	1.829	2.187	3.219	1.243
3	76.20	0.1181	0.9144	9.842	2.743	3.281	4.828	1.864
4	101.60	0.1575	1.219	13.12	3.658	4.374	6.437	2.485
5	127.00	0.1968	1.524	16.40	4.572	5.468	8.047	3.107
6	152.40	0.2362	1.829	19.68	5.486	6.562	9.656	3.728
7	177.80	0.2756	2.134	22.97	6.401	7.655	11.27	4.350
8	203.20	0.3150	2.438	26.25	7.315	8.749	12.87	4.971
9	228.60	0.3543	2.743	29.53	8.230	9.842	14.48	5.592

Example 1 in. = 25.40 mm, 1 m = 3.281 ft., 1 km = 0.6214 mi.

Surface Measure

Unit	Square inches to square centimetres	Square centimetres to square inches	Square feet to square metres	Square metres to square feet	Square yards to square metres	Square metres to square yards	Acres to hectares	Hectares to acres	Square miles to square kilometres	Square kilometres to square miles
1	6.452	0.1550	0.0929	10.76	0.8361	1.196	0.4047	2.471	2.59	0.3861
2	12.90	0.31	0.1859	21.53	1.672	2.392	0.8094	4.942	5.18	0.7722
3	19.356	0.465	0.2787	32.29	2.508	3.588	1.214	7.413	7.77	1.158
4	25.81	0.62	0.3716	43.06	3.345	4.784	1.619	9.884	10.36	1.544
5	32.26	0.775	0.4645	53.82	4.181	5.98	2.023	12.355	12.95	1.931
6	38.71	0.93	0.5574	64.58	5.017	7.176	2.428	14.826	15.54	2.317
7	45.16	1.085	0.6503	75.35	5.853	8.372	2.833	17.297	18.13	2.703
8	51.61	1.24	0.7432	86.11	6.689	9.568	3.237	19.768	20.72	3.089
9	58.08	1.395	0.8361	96.87	7.525	10.764	3.642	22.239	23.31	3.475

Example 1 sq. in. = 6.452 cm² 1 m² = 1.196 sq. yd., 1 sq. mi. = 2.59 km²

Cubic Measure

Unit	Cubic inches to cubic centimetres	Cubic centimetres to cubic inches	Cubic feet to cubic metres	Cubic metres to cubic feet	Cubic yards to cubic metres	Cubic metres to cubic yards	Gallons to cubic feet	Cubic feet to gallons
1	16.39	0.06102	0.02832	35.31	0.7646	1.308	0.1337	7.481
2	32.77	0.1220	0.05663	70.63	1.529	2.616	0.2674	14.96
3	49.16	0.1831	0.08495	105.9	2.294	3.924	0.4010	22.44
4	65.55	0.2441	0.1133	141.3	3.058	5.232	0.5347	29.92
5	81.94	0.3051	0.1416	176.6	3.823	6.540	0.6684	37.40
6	98.32	0.3661	0.1699	211.9	4.587	7.848	0.8021	44.88
7	114.7	0.4272	0.1982	247.2	5.352	9.156	0.9358	52.36
8	131.1	0.4882	0.2265	282.5	6.116	10.46	1.069	59.84
9	147.5	0.5492	0.2549	371.8	6.881	11.77	1.203	67.32

Example 1 cm³ = 0.06102 cu. in., 1 gal. = 0.1337 cu. ft.

Volume or Capacity Measure

Unit	Liquid ounces to cubic centimetres	Cubic centimetres to liquid ounces	Pints to litres	Litres to pints	Quarts to litres	Litres to quarts	Gallons to litres	Litres to gallons	Bushels to hectolitres	Hectolitres to bushels
1	29.57	0.03381	0.4732	2.113	0.9463	1.057	3.785	0.2642	0.3524	2.838
2	59.15	0.06763	0.9463	4.227	1.893	2.113	7.571	0.5284	0.7048	5.676
3	88.72	0.1014	1.420	6.340	2.839	3.785	11.36	0.7925	1.057	8.513
4	118.3	0.1353	1.893	8.454	3.170	4.227	15.14	1.057	1.410	11.35
5	147.9	0.1691	2.366	10.57	4.732	5.284	18.93	1.321	1.762	14.19
6	177.4	0.2029	2.839	12.68	5.678	6.340	22.71	1.585	2.114	17.03
7	207.0	0.2367	3.312	14.79	6.624	7.397	26.50	1.849	2.467	19.86
8	236.6	0.2705	3.785	16.91	7.571	8.454	30.28	2.113	2.819	22.70
9	266.2	0.3043	4.259	19.02	8.517	9.510	34.07	2.378	3.171	25.54

Example 1 L = 2.113 pt., 1 gal. = 3.785 L

TABLE II DECIMAL EQUIVALENTS

DECIMAL EQUIVALENTS					
Fraction	**Decimal Equivalent**		**Fraction**	**Decimal Equivalent**	
	Customary (in.)	Metric (mm)		Customary (in.)	Metric (mm)
1/64 — .015625		0.3969	33/64 — .515625		13.0969
1/32 — .03125		0.7938	17/32 — .53125		13.4938
3/64 — .046875		1.1906	35/64 — .546875		13.8906
1/16 — .0625		1.5875	9/16 — .5625		14.2875
5/64 — .078125		1.9844	37/64 — .578125		14.6844
3/32 — .09375		2.3813	19/32 — .59375		15.0813
7/64 — .109375		2.7781	39/64 — .609375		15.4781
1/8 — .1250		3.1750	5/8 — .6250		15.8750
9/64 — .140625		3.5719	41/64 — .640625		16.2719
5/32 — .15625		3.9688	21/32 — .65625		16.6688
11/64 — .171875		4.3656	43/64 — .671875		17.0656
3/16 — .1875		4.7625	11/16 — .6875		17.4625
13/64 — .203125		5.1594	45/64 — .703125		17.8594
7/32 — .21875		5.5563	23/32 — .71875		18.2563
15/64 — .234375		5.9531	47/64 — .734375		18.6531
1/4 — .250		6.3500	3/4 — .750		19.0500
17/64 — .265625		6.7469	49/64 — .765625		19.4469
9/32 — .28125		7.1438	25/32 — .78125		19.8438
19/64 — .296875		7.5406	51/64 — .796875		20.2406
5/16 — .3125		7.9375	13/16 — .8125		20.6375
21/64 — .328125		8.3384	53/64 — .828125		21.0344
11/32 — .34375		8.7313	27/32 — .84375		21.4313
23/64 — .359375		9.1281	55/64 — .859375		21.8281
3/8 — .3750		9.5250	7/8 — .8750		22.2250
25/64 — .390625		9.9219	57/64 — .890625		22.6219
13/32 — .40625		10.3188	29/32 — .90625		23.0188
27/64 — .421875		10.7156	59/64 — .921875		23.4156
7/16 — .4375		11.1125	15/16 — .9375		23.8125
29/64 — .453125		11.5094	61/64 — .953125		24.2094
15/32 — .46875		11.9063	31/32 — .96875		24.6063
31/64 — .484375		12.3031	63/64 — .984375		25.0031
1/2 — .500		12.7000	1 — 1.000		25.4000

TABLE III POWERS AND ROOTS OF NUMBERS (1 through 100)

Num-ber	Powers		Roots		Num-ber	Powers		Roots	
	Square	Cube	Square	Cube		Square	Cube	Square	Cube
1	1	1	1.000	1.000	51	2,601	132,651	7.141	3.708
2	4	8	1.414	1.260	52	2,704	140,608	7.211	3.733
3	9	27	1.732	1.442	53	2,809	148,877	7.280	3.756
4	16	64	2.000	1.587	54	2,916	157,464	7.348	3.780
5	25	125	2.236	1.710	55	3,025	166,375	7.416	3.803
6	36	216	2.449	1.817	56	3,136	175,616	7.483	3.826
7	49	343	2.646	1.913	57	3,249	185,193	7.550	3.849
8	64	512	2.828	2.000	58	3,364	195,112	7.616	3.871
9	81	729	3.000	2.080	59	3,481	205,379	7.681	3.893
10	100	1,000	3.162	2.154	60	3,600	216,000	7.746	3.915
11	121	1,331	3.317	2.224	61	3,721	226,981	7.810	3.936
12	144	1,728	3.464	2.289	62	3,844	238,328	7.874	3.958
13	169	2,197	3.606	2.351	63	3,969	250,047	7.937	3.979
14	196	2,744	3.742	2.410	64	4,096	262,144	8.000	4.000
15	225	3,375	3.873	2.466	65	4,225	274,625	8.062	4.021
16	256	4,096	4.000	2.520	66	4,356	287,496	8.124	4.041
17	289	4,913	4.123	2.571	67	4,489	300,763	8.185	4.062
18	324	5,832	4.243	2.621	68	4,624	314,432	8.246	4.082
19	361	6,859	4.359	2.668	69	4,761	328,509	8.307	4.102
20	400	8,000	4.472	2.714	70	4,900	343,000	8.367	4.121
21	441	9,261	4.583	2.759	71	5,041	357,911	8.426	4.141
22	484	10,648	4.690	2.802	72	5,184	373,248	8.485	4.160
23	529	12,167	4.796	2.844	73	5,329	389,017	8.544	4.179
24	576	13,824	4.899	2.884	74	5,476	405,224	8.602	4.198
25	625	15,625	5.000	2.924	75	5,625	421,875	8.660	4.217
26	676	17,576	5.099	2.962	76	5,776	438,976	8.718	4.236
27	729	19,683	5.196	3.000	77	5,929	456,533	8.775	4.254
28	784	21,952	5.292	3.037	78	6,084	474,552	8.832	4.273
29	841	24,389	5.385	3.072	79	6,241	493,039	8.888	4.291
30	900	27,000	5.477	3.107	80	6,400	512,000	8.944	4.309
31	961	29,791	5.568	3.141	81	6,561	531,441	9.000	4.327
32	1,024	32,798	5.657	3.175	82	6,724	551,368	9.055	4.344
33	1,089	35,937	5.745	3.208	83	6,889	571,787	9.110	4.362
34	1,156	39,304	5.831	3.240	84	7,056	592,704	9.165	4.380
35	1,225	42,875	5.916	3.271	85	7,225	614,125	9.220	4.397
36	1,296	46,656	6.000	3.302	86	7,396	636,056	9.274	4.414
37	1,369	50,653	6.083	3.332	87	7,569	658,503	9.327	4.481
38	1,444	54,872	6.164	3.362	88	7,744	681,472	9.381	4.448
39	1,521	59,319	6.245	3.391	89	7,921	704,969	9.434	4.465
40	1,600	64,000	6.325	3.420	90	8,100	729,000	9.487	4.481
41	1,681	68,921	6.403	3.448	91	8,281	753,571	9.539	4.498
42	1,764	74,088	6.481	3.476	92	8,464	778,688	9.592	4.514
43	1,849	79,507	6.557	3.503	93	8,649	804,357	9.644	4.531
44	1,936	85,184	6.633	3.530	94	8,836	830,584	9.695	4.547
45	2,025	91,125	6.708	3.557	95	9,025	857,375	9.747	4.563
46	2,116	97,336	6.782	3.583	96	9,216	884,736	9.798	4.579
47	2,209	103,823	6.856	3.609	97	9,409	912,673	9.849	4.595
48	2,304	110,592	6.928	3.634	98	9,604	941,192	9.900	4.610
49	2,401	117,649	7.000	3.659	99	9,801	970,299	9.950	4.626
50	2,500	125,000	7.071	3.684	100	10,000	1,000,000	10.000	4.642

TABLE IV

CIRCUMFERENCES AND AREAS (0.2 to 9.8; 10 to 99)

Diameter	Circum.	Area	Diameter	Circum.	Area
0.2	0.628	0.0314	31	97.39	754.8
0.4	1.26	0.1256	32	100.5	804.2
0.6	1.88	0.2827	33	103.7	855.3
0.8	2.51	0.5026	34	106.8	907.9
1	3.14	0.7854	35	110	962.1
1.2	3.77	1.131	36	113.1	1,017.9
1.4	4.39	1.539	37	116.2	1,075.2
1.6	5.02	2.011	38	119.4	1,134.1
1.8	5.65	2.545	39	122.5	1,194.6
2	6.28	3.142	40	125.7	1,256.6
2.2	6.91	3.801	41	128.8	1,320.3
2.4	7.53	4.524	42	131.9	1,385.4
2.6	8.16	5.309	43	135.1	1,452.2
2.8	8.79	6.158	44	138.2	1,520.5
3	9.42	7.069	45	141.4	1,590.4
3.2	10.05	7.548	46	144.5	1,661.9
3.4	10.68	8.553	47	147.7	1,734.9
3.6	11.3	10.18	48	150.8	1,809.6
3.8	11.93	11.34	49	153.9	1,885.7
4	12.57	12.57	50	157.1	1,963.5
4.2	13.19	13.85	51	160.2	2,042.8
4.4	13.82	15.21	52	163.4	2,123.7
4.6	14.45	16.62	53	166.5	2,206.2
4.8	15.08	18.1	54	169.6	2,290.2
5	15.7	19.63	55	172.8	2,375.8
5.2	16.33	21.24	56	175.9	2,463
5.4	16.96	22.9	57	179.1	2,551.8
5.6	17.59	24.63	58	182.2	2,642.1
5.8	18.22	26.42	59	185.4	2,734
6	18.84	28.27	60	188.5	2,827.4
6.2	19.47	30.19	61	191.6	2,922.5
6.4	20.1	32.17	62	194.8	3,019.1
6.6	20.73	34.21	63	197.9	3,117.3
6.8	21.36	36.32	64	201.1	3,217
7	21.99	38.48	65	204.2	3,318.3
7.2	22.61	40.72	66	207.3	3,421.2
7.4	23.24	43.01	67	210.5	3,525.7
7.6	23.87	45.36	68	213.6	3,631.7
7.8	24.5	47.78	69	216.8	3,739.3
8	25.13	50.27	70	219.9	3,848.5
8.2	25.76	52.81	71	223.1	3,959.2
8.4	26.38	55.42	72	226.2	4,071.5
8.6	27.01	58.09	73	229.3	4,185.4
8.8	27.64	60.82	74	232.5	4,300.8
9	28.27	63.62	75	235.6	4,417.9
9.2	28.9	66.48	76	238.8	4,536.5
9.4	29.53	69.4	77	241.9	4,656.6
9.6	30.15	72.38	78	245	4,778.4
9.8	30.78	75.43	79	248.2	4,901.7
10	31.41	78.54	80	251.3	5,026.6
11	34.55	95.03	81	254.5	5,153
12	37.69	113	82	257.6	5,281
13	40.84	132.7	83	260.8	5,410.6
14	43.98	153.9	84	263.9	5,541.8
15	47.12	176.7	85	267.0	5,674.5
16	50.26	201	86	270.2	5,808.8
17	53.4	226.9	87	273.3	5,944.7
18	56.54	254.4	88	276.5	6,082.1
19	59.69	283.5	89	279.6	6,221.2
20	62.83	314.1	90	282.7	6,361.7
21	65.97	346.3	91	285.9	6,503.9
22	69.11	380.1	92	289.0	6,647.6
23	72.25	415.4	93	292.2	6,792.9
24	75.39	452.3	94	295.2	6,939.8
25	78.54	490.8	95	298.5	7,088.2
26	81.68	530.9	96	301.6	7,238.2
27	84.82	572.5	97	304.7	7,389.8
28	87.96	615.7	98	307.9	7,543.0
29	91.1	660.5	99	311.9	7,697.7
30	94.24	706.8			

TABLE V

TAP DRILL SIZES FOR AMERICAN NATIONAL FORM THREADS
(CUSTOMARY AND METRIC DRILL SIZES)

Diam. of Thread	Threads per Inch	Drill	Decimal Equiv.	Diam. of Thread	Threads per Inch	Drill	Decimal Equiv.
No. 0—.060	80 NF	3/64	0.0469		12 N	39/64	0.6094
1—.073	64 NC	1.5 mm	0.0591	11/16	24 NEF	16.5 mm	0.6496
	72 NF	53	0.0595		10 NC	16.5 mm	0.6496
2—.086	56 NC	50	0.0700		12 N	17 mm	0.6693
	64 NF	50	0.0700	3/4	16 NF	17.5 mm	0.6890
3—.099	48 NC	5/64	0.0781		20 NEF	45/64	0.7031
	56 NF	45	0.0820		12 N	18.5 mm	0.7283
4—.112	40 NC	43	0.0890	13/16	16 N	3/4	0.7500
	48 NF	42	0.0935		20 NEF	49/64	0.7656
5—.125	40 NC	38	0.1015		9 NC	49/64	0.7656
	44 NF	37	0.1040		12 N	20 mm	0.7874
6—.138	32 NC	36	0.1065	7/8	14NF	20.5 mm	0.8071
	40 NF	33	0.1130		16 N	13/16	0.8125
8—.164	32 NC	29	0.1360		20 NEF	21 mm	0.8268
	36 NF	29	0.1360		12 N	55/64	0.8594
10—.190	24 NC	25	0.1495	15/16	16 N	7/8	0.8750
	32 NF	21	0.1590		20 NEF	22.5 mm	0.8858
12—.216	24 NC	16	0.1770		8 NC	7/8	0.8750
	28 NF	14	0.1820		12 N	59/64	0.9219
1/4	20 NC	7	0.2010	1	14 NF	23.5 mm	0.9252
	28 NF	3	0.2130		16 N	15/16	0.9375
	32 NEF	7/32	0.2188		20 NEF	61/64	0.9531
5/16	18 NC	F	0.2570		6 NC	1 21/64	1.3281
	24 NF	I	0.2720		8N	1 3/8	1.3750
	32 NEF	9/32	0.2812	1 1/2	12 NF	36 mm	1.4173
3/8	16 NC	5/16	0.3125		16N	1 7/16	1.4375
	24 NF	Q	0.3320		18 NEF	1 29/64	1.4531
	32 NEF	11/32	0.3438		4 1/2 NC	1 25/32	1.7812
7/16	14 NC	U	0.3680	2	8 N	1 7/8	1.8750
	20 NF	25/64	0.3906		12 N	1 59/64	1.9219
	28 NEF	Y	0.4040		16 NEF	1 15/16	1.9375
	12 N	27/64	0.4219		4 NC	2 1/4	2.2500
1/2	13 NC	27/64	0.4219	2 1/2	8 N	2 3/8	2.3750
	20 NF	29/64	0.4531		12 N	61.5 mm	2.4213
	28 NEF	15/32	0.4687		16 N	2 7/16	2.4375
	12 NC	31/64	0.4844		4 NC	2 3/4	2.7500
9/16	18 NF	33/64	0.5156	3	8 N	2 7/8	2.8750
	24 NEF	33/64	0.5156		12 N	74 mm	2.9134
	11 NC	17/32	0.5312		16 N	2 15/16	2.9375
5/8	12 N	35/64	0.5469				
	18 NF	14.5 mm	0.5709				
	24 NEF	37/64	0.5781				

Glossary

Acute Angle - An angle of less than 90 degrees.

Alloy - Two or more metals in combination.

Allowance - Minimum clearance or maximum interference between mating parts.

Angle - A space formed by the intersection of two lines or planes.

Arc - Part of a circle.

Area - The amount of surface contained within a figure.

Architect - A person who designs and oversees the construction of buildings.

Architectural Drafter - A person who prepares all types of architectural drawings and documents.

Arrow Heads - Points used on the end of dimension lines or leaders.

Assembly Drawing - A drawing showing all the parts in their proper locations.

Basic Size - The size of an object from which tolerances are applied.

Bearing - A supporting member for a rotating shaft.

Bevel - An inclined edge, not a right angle to a joining surface.

Bisect - To divide into two equal parts.

Blue Print - A reproduction of a drawing having a bright blue background and white lines.

Bolt Circle - A circular center line on which holes are located.

Bore - To enlarge a hole with a boring bar or other machine tool.

Cam - A rotating member used to change circular motion to reciprocating motion.

Casting - An object made by pouring molten metal into a mold.

Center Line - A thin line representing the center of holes or the axis of cylindrical objects.

Chamfer - A slight bevel on the end of a shaft or corner of an object to avoid a sharp edge.

Circumference - The perimeter of a circle.

Circumscribe - To draw a figure outside another figure.

Civil Drafter - A person who prepares drawings dealing with highways, bridges, tunnels, etc.

Clearance Fit - Class of fit where clearance is always maintained between mating parts.

Collar - A round flange or ring fitted to a shaft to prevent sliding.

Computer-Aided Drafting (CAD) - The use of a computer as a tool for drawing objects.

Concave - A curved depression in the surface of an object.

Concentric - Having the same center.

Cone - A geometric shape that tapers uniformly from a point to a circular base.

Conical - Shaped like a cone.

Convex - An exterior rounded surface of an object.

Core - To form a hollow area in a part to be cast in a mold.

Counterbore - To enlarge the end of a hole cylindrically to a given depth.

Countersink - To enlarge the end of a hole conically to a specified angle.

Crosshatching - Lines added to a sectional drawing to indicate the material of the part.

Cylinder - A geometric figure having a uniform circular cross-section for its entire length.

Detail Drawing - The drawing of a single part including its exact dimensions.

Development - A drawing of the surface of an object unfolded or rolled out on a plane.

Diagonal - A line running across a figure from opposite corners.

Diameter - The distance across a circle passing through its center.

Dimension Line - A line drawn between extension lines that contains a size dimension.

Dowel - A cylindrical pin used primarily to prevent sliding between two contacting surfaces.

Drafter - A person who works with original designs to make production drawings.

Drill - To cut a cylindrical hole using a drill bit.

Eccentric - Not having a common center.

Equilateral Triangle - A triangle having all sides and angles equal.

Ellipse - An oval circular figure with two centers and a major and minor axis.

Engineer - A person who solves technical problems for a company.

Entity - A point, line, arc, circle, or text used to generate CAD drawings.

Erasing Shield - A drafting aid used to protect specific features when erasing.

Extension Lines - Thin lines drawn from object lines to indicate where dimensions begin and end.

Fillet - The inside rounded corner on a casting.

Fit - The degree of looseness or tightness between two mating parts.

Flange - A relatively thin rim around a part.

Floor Plan - The top view of a building at a specified floor level.

Full Section - A view of an object that is cut in half by a cutting-plane with the front half removed.

Galvanize - To cover a surface with a molten alloy to prevent rusting.

Gasket - A thin piece of material placed between two surfaces to make a tight joint.

Graphic Symbols - Symbolic representations used to simplify complicated items.

Half Section - A view of an object in which one quarter of the object has been removed.

Hidden Line - A thin line made up of a series of short dashes showing a surface that is hidden behind other surfaces.

Hexagon - A six-sided geometric shape.

Inclined - Making an angle with another line or plane.

Inscribe - To draw a figure inside another figure.

Interference Fit - Class of fit where interference is always maintained between mating parts.

Isometric Drawing - A pictorial drawing showing three connected planes of an object represented in two dimensions.

Isosceles Triangle - A triangle having two sides and two base angles equal.

Key - A small piece of metal fitting in both the shaft and hub to prevent circumferential motion.

Keyway - A slot in a hub or portion surrounding a shaft to receive a key.

Lathe - A machine used to shape material by rotation against a tool.

Leader - A line drawn from a note to the object where the note applies.

Level - Transparent layer where entities are generated and recorded on CAD drawings.

Limit Dimensioning - Dimensioning system where class of fits are controlled by applying tolerances and allowances.

Mechanical Drafting - The drafting of drawings dealing with primarily mechanical objects.

Micrometer - An instrument used to measure thickness (diameter) with precision.

Nominal Size - The specified size of an object which may not be its actual size.

Oblique Angle - An angle that does not contain 90 degrees.

Oblique Drawing - A pictorial drawing showing one face parallel to the viewer.

Obtuse Angle - An angle containing more than 90 degrees.

Octagon - An eight-sided geometric shape.

Orthographic Projection - The method used to represent two and three dimensional objects using standard views.

Parallel - Two or more lines or planes that are always the same distance apart.

Pattern - A model used in forming a mold for a casting.

Pentagon - A five-sided geometric shape.

Perimeter - The distance around the outside of a geometric figure.

Perpendicular - A line or plane 90 degrees to another line or plane; at right angles.

Perspective Drawing - A pictorial drawing that closely approximates the actual optical image of an object.

Pitch - The distance from one point on a screw thread to the same point on the next thread.

Plot Plan - A map or drawing of an area that shows boundaries of lots and other parcels of property.

Polygon - Plane figure bounded by straight lines.

Protractor - A drawing device used to measure angles.

Radius - One half the diameter of a circle.

Ream - To enlarge a hole slightly to give it greater accuracy.

Rectangle - A geometric shape having four straight sides and four 90 degree angles.

Right Angle - An angle containing 90 degrees and formed by two perpendicular lines.

Rivet - A small fastener used to join objects together.

Round - The outside rounded corner on a casting.

Scale - A measuring device used to lay out or measure distances.

Scalene Triangle - A triangle having three unequal sides.

Schematic - A diagram usually containing electric or electronic circuitry and symbols.

Sectional View - A view of an object showing interior detail.

Shaft - A revolving bar usually cylindrical serving to transmit motion.

Shim - A thin piece of material used as a spacer in adjusting two parts.

Spotface - To produce a shallow circular bearing surface beneath the surface of a part.

Structural Drafter - A person who prepares drawings dealing with the structural portion of buildings.

Surveyor - A person skilled in land measurement.

Tangent - The point at which a straight line and curved lines meet.

Tap - The tool used to cut internal threads.

Tap Drill - The drill used to make a hole for an internal thread.

Template - A shape that is used as a pattern or guide.

Tolerance - Permissible variance in the size of a part.

Trapezoid - A plane figure having two parallel and two nonparallel sides.

Tread - The step or horizontal member of a stair.

Triangle - A three-sided geometric figure.

Triangulation - A method of development using a system of imaginary triangles.

Truncate - To cut off a geometrical solid at an angle to its base.

Vellum - Semi-translucent drawing paper used to make whiteprints.

Vernier Scale - A graduated scale used to obtain very precise measurements.

Volume - The amount of space occupied by an object measured in cubic units.

Weld - Uniting metal parts by pressure or fusion welding processes.

Whiteprint - A reproduction of a drawing having a light colored background and bluish lines.

Working Drawing - A drawing containing the necessary information that will allow someone to work directly from it.

ANSWERS TO ODD-NUMBERED PROBLEMS

SECTION 1 WHOLE NUMBERS

UNIT 1 ADDITION OF WHOLE NUMBERS

1. 227 in.
3. 791 mm
5. 155 in.
7. 221 pencils
9. 563 mm

11. a. 74 mm
     b. 51 mm
13. 61 ft.
15. A = 24 in.
     B = 20 in.
17. 1,900 min.

19. 44 details
21. 96 keys
23. 148 doors
25. a. 5
     b. 8
     c. 3
     d. 6

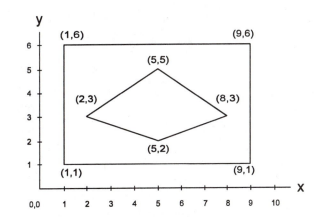

27. a. 2,795 ft.
     b. 376 ft.
     c. 1,936 ft.

29. Length = 22 ft.
     Height = 60 ft.

UNIT 2 SUBTRACTION OF WHOLE NUMBERS

1. 82 ft.
3. 141 mm
5. 315 yd.
7. 154 ft.
9. 139 days
11. 1,377 entities
13. 24 parts

15. 63 drafters
17. A = 52 mm
     B = 135 mm
     C = 70 mm
19. A = 21 mm
     B = 27 mm
     C = 12 mm

21. 66 symbols
23. 170 ft.
25. 148 ft.
27. 7
29. 8 ft.

UNIT 3 MULTIPLICATION OF WHOLE NUMBERS

1. 518 cm
3. 4,140 in.
5. 221 ft.
7. 864 yd.
9. 24 in.
11. 56 hr.
13. 216 mm
15. 102 lb.
17. 1,900 cm

19. 210°
21. 256 mm
23. 28,048 bits
25. A = 232
    B = 64
    C = 208
    D = 248
    E = 56
    Length = 576

27. Height = 35
    Width = 49
    Diameter = 14
29. A = 288
    B = 153
    C = 189
    D = 387
    E = 594

UNIT 4 DIVISION OF WHOLE NUMBERS

1. 64 in.
3. 63 cm
5. 36 mm
7. 6 sections
9. 12 line segments
11. 83 hr.

13. 7 lb.
15. $4.00
17. 67 mm
19. 60 packages
21. 4,789 bytes

23. Height = 16 in.
    Width = 24 in.
25. A = 7 mm
    B = 6 mm
27. 15 ft.
29. 30 active levels

UNIT 5 COMBINED OPERATIONS OF WHOLE NUMBERS

1. $\frac{(16 - 6)}{2} = 5"$
3. 194 mm
5. a. 1,275 T squares
   b. 1,227 T squares
7. 117 sq. in.
9. 5,171 sq. in.
11. 82 mm
13. 24 engineers
15. 6 in.
17. 4,560 mm

19. 133 sq. in.
21. a. 194 levels
     assigned
   b. 62 levels not
     assigned
23. 53,868 bytes
25. Height = 7 ft.
    Width = 12 ft.
27. A = 144
    B = 104
    C = 112

29. AB = 73 ft.
    BC = 23 ft.
    FG = 59 ft.
    GH = 47 ft.
    HI = 91 ft.
    AI = 58 ft.
31. A = 13
    B = 31
    C = 42

SECTION 2 COMMON FRACTIONS

UNIT 6 ADDITION OF COMMON FRACTIONS

1. 1 in.
3. $\frac{3}{4}$ in.
5. $10\frac{1}{16}$ in.
7. $8\frac{2}{3}$ hr.
9. $12\frac{5}{12}$ yd.
11. $3\frac{3}{16}$ in.

13. $8\frac{3}{4}$ in.
15. A = $3\frac{1}{2}$ in.
    B = $3\frac{13}{16}$ in.
17. $2\frac{15}{16}$ in.
19. A = $4\frac{23}{64}$ in.
    B = $7\frac{25}{64}$ in.

21. $7\frac{5}{32}$ in.
23. $7\frac{3}{32}$ in.
25. A = $17\frac{11}{32}$ in.
    B = $8\frac{3}{4}$ in.
27. $30\frac{1}{4}$ in.
29. $6\frac{9}{32}$ in.

UNIT 7 SUBTRACTION OF COMMON FRACTIONS

1. in.
3. $1\frac{9}{16}$ yd.
5. $\frac{45}{64}$ in.
7. $1\frac{7}{12}$ hr.
9. $\frac{5}{8}$ ft.
11. $\frac{15}{64}$ in.
13. $18\frac{1}{4}$ hr.

15. $9\frac{1}{8}$ in.
17. $\frac{7}{16}$ in.
19. $\frac{9}{16}$ in.
21. a. $2\frac{13}{16}$ in.
    b. $1\frac{5}{64}$ in.
23. $4\frac{5}{64}$ in.

25. A = $1\frac{7}{16}$ in.
    B = $\frac{7}{16}$ in.
27. A = $1\frac{49}{64}$ in.
    B = $1\frac{1}{16}$ in.
29. A = $2\frac{59}{64}$ in.
    B = 2 in.
    C = $\frac{15}{16}$ in.

UNIT 8 MULTIPLICATION OF COMMON FRACTIONS

1. $\frac{3}{16}$
3. $\frac{7}{64}$
5. $\frac{13}{24}$
7. 20
9. $3\frac{33}{64}$
11. 42 in.
13. $2\frac{1}{4}$ in.

15. 645 cents or $6.45
17. Floors = $101\frac{3}{4}$ in.
    A = 132 in.
19. $4\frac{1}{2}$ in.
21. $8\frac{1}{4}$ in.
23. $3\frac{3}{4}$ in.

25. $18\frac{3}{8}$ in.
27. A = $7\frac{1}{2}$ in.
    B = $6\frac{1}{4}$ in.
    C = $12\frac{1}{2}$ in.
29. A = $52\frac{23}{32}$ in.
    B = $21\frac{7}{8}$ in.

UNIT 9 DIVISION OF COMMON FRACTIONS

1. 4
3. $20\frac{2}{5}$
5. $6\frac{1}{2}$ in.
7. $\frac{9}{98}$
9. $\frac{3}{7}$
11. $1\frac{27}{64}$ in.

13. $\frac{13}{16}$ in.
15. 12 jobs
17. 160 pins
19. 15 erasers
21. $\frac{7}{16}$ in.

23. a. $25\frac{1}{2}$ in.
    b. $27\frac{1}{4}$ in.
25. $\frac{7}{8}$ in.
27. $A = 1\frac{29}{32}$ in.
    $B = \frac{25}{32}$ in.
29. $A = \frac{3}{16}$ in.
    $B = \frac{7}{32}$ in.

UNIT 10 MULTIPLE OPERATIONS OF COMMON FRACTIONS

1. $2\frac{3}{128}$ in.
3. $10\frac{1}{3}$
5. 5
7. Aluminum is thicker.
9. $4\frac{1}{4}$ hr.
11. 80 shields
13. $6\frac{3}{32}$ in.
15. 31 ft.
17. 44 handles
19. $8\frac{15}{16}$ in.

21. Cabin = 85 ft. 0 in.
    Deck = 42 ft. 0 in.
    Difference = 43 ft. 0 in.
23. a. 9 in.
    b. 11 in.
    c. $2\frac{1}{2}$ in.
    d. $3\frac{1}{4}$ in.
25. $A = 1\frac{17}{32}$ in.
    $B = 5\frac{57}{64}$ in.
    $C = 2\frac{1}{32}$ in.

    $D = 2\frac{7}{16}$ in.
    $E = 2\frac{1}{8}$ in.
27. $A = 4$ ft. 6 in.
    $B = 3$ ft. 6 in.
    $C = 8$ ft. 6 in.
    $D = 7$ ft. 0 in.
    Perimeter = 67 ft. 0 in.
29. $A = 1\frac{9}{64}$ in.
    Diameter 1 = $3\frac{1}{8}$ in.
    Diameter 2 = $13\frac{3}{4}$ in.

SECTION 3 DECIMAL FRACTIONS

UNIT 11 ADDITION OF DECIMAL FRACTIONS

1. 0.887
3. 0.8768
5. 1.2605
7. 45.346
9. 200.72 mm
11. $21.64

13. 40.75 hr.
15. $82.85
17. 23.57 cm
19. $163.86
21. 429.05 hrs.

23. $7,586.70
25. 6.918 in.
27. 7.474 in.
29. Height = 6 in.
    Width = 4.8244 in.

UNIT 12 SUBTRACTION OF DECIMAL FRACTIONS

1. 0.59
3. 1.062
5. 7.71
7. 40.817
9. 7.27 mm
11. 17.98 mm

13. $60.56
15. 10.48 lb.
17. 75.065 cm
19. 0.0157 in.
21. 1.463 in.
23. $2,661.70

25. 18.3297 in.
27. .742 in.
29. Outer = 45.794 in.
    Sum-Inner = 39.512 in.
    Difference = 6.282 in.

UNIT 13 MULTIPLICATION OF DECIMAL FRACTIONS

1. 0.0498
3. 0.40455
5. 3.6828
7. 22.2042
9. 3 in.
11. 2.4375 in.
13. 30.336 lb.

15. $452.16
17. 0.91875 in.
19. 14.4375 in.
21. 26.70 in.
23. Height = 13.04 cm
    Width = 25.28 cm
    Depth = 12.64 cm

25. Height = 1.875 in.
    Width = 1.875 in.
27. Height = 18.375 in.
    Width = 15.75 in.
29. Height = 1.75 in.
    Width = 2.625 in.

UNIT 14 DIVISION OF DECIMAL FRACTIONS

1. 0.24
3. 7.0
5. 0.0003
7. 29.31
9. 24 dividers

11. 35.18 mm
13. 0.036 in.
15. $181.50
17. $937.25
19. 1.473 in.

21. 15 levels
23. $21.75 per hour
25. 3 inch diameter
27. $1.87 each
29. Average of 16.06 hrs.

UNIT 15 DECIMAL AND COMMON FRACTION EQUIVALENTS

Answers are rounded to the nearest thousandth.

1. 0.286
3. 0.542
5. $\frac{3}{16}$
7. $\frac{13}{16}$
9. 2.438 in.
11. 28$\frac{3}{4}$ lb.

13. 2 $\frac{33}{64}$ in.
15. 4 in.
17. A = 1$\frac{3}{16}$ in.
    B = $\frac{13}{32}$ in.
19. $\frac{3}{32}$ in.
21. A = 125 in.
    B = $\frac{1}{16}$ in.

23. 1.75 in.
25. $\frac{9}{32}$ in.
27. 1$\frac{1}{2}$ in.
29. A = 5$\frac{1}{2}$ in.
    B = 3$\frac{3}{32}$ in.

UNIT 16 COMBINED OPERATIONS WITH DECIMAL FRACTIONS

Answers are rounded to the nearest thousandth.

1. 0.176
3. 12.199
5. 273.058
7. 0.723 in.
9. A = 27.30 m
    B = 150.150 m
11. 50.24 cm

13. A = 7.151 in. (Rounded)
    B = 1.287 in. (Rounded)
15. a. $25.54
    b. 105 pencils
17. Overall length = 13.27 in.
    A = 505 in.
19. A = 1.85 in.
    B = 1.45 in.

21. a. $7,522.80
    b. $6,874.87
    c. $37,614.00
23. 3 $\frac{5}{64}$ in.
25. Height = 192.1 in.
    Width = 239.1 in.
27. 6.469 sq. in.
29. 38.0382 in.

SECTION 4 PERCENT, AVERAGES, AND ESTIMATING

UNIT 17 PERCENT AND PERCENTAGE

1. 3.36
3. 11.25
5. 250.32
7. 33
9. 139 $\frac{1}{2}$ hr.
11. 17
13. $3,500

15. 15 min.
17. 345
19. 1.414 inches
21. 3.569"
23. 6 CAD operators
    % Structural = 16%
    % Mechanical = 25%

% Civil = 18%
% Architectural = 33%
% CAD = 8%
25. Height = 4.944"
    Width = 2.588"

UNIT 18 INTEREST AND DISCOUNTS

1. $4,320.00
3. 15%
5. Cost = .07 cent sheet
    Discount = $195.30
7. $2.73 each
9. 1,086
11. $804.88

13. Materials = 1%
    Labor = 76%
    Overhead = 23%
15. a. $267.53
    b. $414.24
    c. $181.23
17. a. $513.60
    b. $433.89
    c. $18.08

19. $7,695
21. $303.75
23. $1,368.00
25. $12,470.00
27. 15 $\frac{1}{4}$ %
29. $111.75

UNIT 19 AVERAGES

1. $11.58
3. 37.20 mi.
5. 83

7. $^{17}/_{24}$
9. $560.33
11. 15

13. $45.94
15. 6.29 lb.

UNIT 20 ESTIMATING

1. 91.1%
3. 26%

5. 6.2%
7. 11%

9. 34.3%

UNIT 21 TOLERANCES

1. .029 in.
3. Upper limit = 4.390
   Lower limit = 4.366
   Tolerance = .024
5. 2.566

7. 5.142
9. Hole = .0005
   Shaft = .0007
11. Hole tolerance = .0005
   Shaft tolerance = .0004

Allowance = -.0012
Type of fit = Interference
13. -.016
15. <u>1.357</u>
    1.368

SECTION 5 MEASUREMENT

UNIT 22 LINEAR MEASURE

1. a. 24 in.
   b. 45 in.
   c. 55¾ in.
   d. 63 ¼ in.
3. a. 42 in.
   b. 71.25 in.
   c. 88.44 in.
   d. 116.38 in.
5. A = $^{7}/_{64}$ in.

B = $^{35}/_{64}$ in.
C = $^{61}/_{64}$ in.
D = 1 $^{13}/_{64}$ in.
E = $^{9}/_{32}$ in.
F = $^{25}/_{32}$ in.
7. A = 12 mm
   B = 23 mm
   C = 36 mm
   D = 47 mm

E = 59 mm
F = 74 mm
9. a. 3.376 in.
   b. 2.641 in.
   c. 2.021 in.
11. A = 0' – 2"
    B = 1' – 4"
    C = 1' – 8"
    D = 2' – 10"

294 *Answers to Odd-Numbered Problems*

UNIT 22 LINEAR MEASURE (CONT.)

13. a. 16' – 2"
    b. 18' – 7"
    c. 19' – 3"
    d. 7' – 5"
    e. 10' – 10 ½ "

    f. 8' – 11"
    g. 9' – 9 ½ "
    h. 8' – 7 ½ "
    i. 1' – 9 ½ "
    j. 2' – 1 ½ "

    k. 5' – 1 ¼ "
    l. 1' – 8 ½ "
    m. 1' – 7 ⅝ "
    n. 0' – 3 ½ "

UNIT 23 AREA MEASURE

1. 432 sq. in.
3. 56.70 sq. in.
5. 7,169 mm^2
7. 10.563 sq. in.
9. 7,290 mm^2
11. 48 lb.
13. 184.04 sq. in.
15. 7.5 lb.

17. 9.695 sq. in.
19. 306.51 sq. in.
21. DIA = 15.47 in.
23. 1.069 sq. in.
25. 279.45 cm^2
27. 0.684 lb.
29. 13.364 sq. in.
31. 25.608 sq. in.

33. 21.772 sq. in.
35. 25.967 sq. in.
37. 33.635 cm^2
39. 1,387.877 dm^2
41. 557.75 sq. ft.
43. 19.8125 sq. in.
45. 6.1359 sq. in.

UNIT 24 VOLUME MEASURE

1. 6,912 cu. in.
3. 175 cu. ft.
5. 2,114 cu. ft.
7. 3,990 mm^3
9. 81.844 cu. in.
11. 13.54 in.
13. 19.10 lb.
15. 896 cu. yd.

17. ¼ ft.
19. 788 ft.^{3}
21. 14.137 cu. in.
23. 2.105 cu. in.
25. 109.08 cu. yd.
27. 5,640.19 gal.
29. 0.737 cu. in.
31. 24,480 gal.

33. 6,446.563 mm^3
35. 96.45 lb.
37. 53.041 cu. in.
39. 4.44 lb.
41. 13.2014 cu. in.
43. 15.70 cu. in.
45. 40.98 cu. in.

UNIT 25 EQUIVALENT MEASUREMENT UNITS AND CONVERSION

Answers may vary when The Table of Equivalent English and Metric Units of Measure is used from the Appendix.

1. 152.400 mm
3. 41.91 cm
5. 4.267 m
7. 156.162 in.
9. 66.929 in.

11. A = 42.545 mm
    B = 28.626 mm
    C = 12.700 mm
    D = 106.363 mm
13. 228.600 mm

15. 17 pieces each 7 in.
17. 1.339 in.
19. 4.503 in.
21. 30.48 cm
23. 294.5 sq. in.

UNIT 25 EQUIVALENT MEASUREMENT UNITS AND CONVERSION (CONT.)

25. 12.37 cm^2
27. 705.64 mm^2
29. 65.40 m^2
31. 19.98 sq. yd.
33. 1.805 sq. in.
35. 298.451 sq. in.
37. 80.849 sq. yd.

39. 300.646 sq. ft.
41. 69.59 sq. in.
43. 13.76 m^3
45. 939.75 cu. in.
47. 1647.62 cu. in.
49. 11.77 cu. yd.
51. 14.25 cu. yd.

53. 2.368 cu. ft.
55. Spacer = 23.15 cu. in.
57. 4.16 dm^3
59. 83.35 m^3
61. 41.44 in.^3

UNIT 26 ANGULAR MEASURE

1. 180°
3. 270°
5. 26° 10'
7. 57° 45'
9. 40°
11. 180°
13. 17 degrees
15. 33° 36' 20"

17. 65° 31' 23"
19. ∠A = 90°
    ∠B = 90°
    ∠C = 15°
19. ∠D = 120°
    ∠E = 30°
    ∠F = 90°
21. 98° 13' 36"

23. A = 105°
    B = 143°
    C = 120°
    D = 90°
25. C = 78° 58'
    D = 22° 42'

UNIT 27 SCALED MEASUREMENT

1. a = 4¾ in.
   b = 3¼ in.
3. a = 2¹¹⁄₁₆ in.
   b = 4⁹⁄₁₆ in.
5. a = 1⁴³⁄₆₄ in.
   b = 3⁵³⁄₆₄ in.
7. a = 4²⁵⁄₃₂ in.
   b = 2¹¹⁄₃₂ in.
9. A = 3 in.
   B = 5 in.
   C = 2½ in.
   D = 9¼ in.

E = 19¾ in.
F = 6 in.
11. A = 19 mm
    B = 23 mm
    C = 28 mm
    D = 32 mm
    E = 93 mm
    F = 44 mm
    G = 16 mm
    H = 13 mm
13. A = ⅞ in.
    B = ⅜ in.

C = 3⅛ in.
D = 4⅛ in.
E = 1⅞ in.
F = 1⅜ in.
15. A = 320'
    B = 320'
    C = 125'
    D = 340'
    E = 260'
    F = 110'
    G = 160'

SECTION 6 RATIO AND PROPORTION

UNIT 28 RATIO

1. $\frac{15}{8}$
3. $\frac{6}{1}$
5. $\frac{60}{12} = \frac{5}{1}$
7. $\frac{1}{16}$
9. $\frac{32}{1}$
11. 4:1

13. A = 1 $\frac{1}{16}$ in.
     B = $\frac{11}{16}$ in.
     C = $\frac{7}{16}$ in.
     D = 1 $\frac{11}{32}$ in.
15. 24:1
17. A = .90 in.
     B = .60 in.

C = .50 in.
D = .15 in.
E = .20 in.
F = .30 in.
G = .40 in.
H = .25 in.
19. 1:48

UNIT 29 PROPORTION

1. x = 12 cm
3. Y = 20.25
5. x = 50 rpm
7. x = 19 in.

9. x = 136 min.
11. x = 25 men
13. x = 26 teeth

15. x = 3.6
17. 4.25 or 4 $\frac{1}{4}$ in.
19. x = 25 rivets

SECTION 7 APPLIED ALGEBRA

UNIT 30 SYMBOLS AND EQUATIONS

1. 4
3. 3.77
5. 35
7. x = 4
9. Z = 5 $\frac{1}{2}$
11. L = 6D

13. A = 7 $\frac{1}{8}$ L
     B = 6 $\frac{13}{16}$ L
15. A = 10 $\frac{7}{16}$ x
     B = 8 $\frac{5}{8}$ x
     C = 14 $\frac{1}{4}$ x

17. 7.503 in.
19. 41 $\frac{9}{16}$ in.
21. 8.75

UNIT 31 POWERS AND ROOTS

1. 16
3. 81
5. 343
7. 2,197
9. 15.625

11. 0.125
13. 7.00
15. 2.09
17. 43.57
19. 1.29

21. 14.75
23. 1.767 sq. in.
25. 15.811 mm
27. 7.595 in.
29. 6.5

UNIT 32 FORMULAS AND HANDBOOK DATA

1. 0.97815
3. 0.51504
5. 84°
7. a. 6 in.
   b. 6.5 in.
   c. 0.25 in.
   d. 0.393 in.
   e. 0.539 in.

9. 7.2 in.
11. 21.99 mm
13. a. 1.750 in.
    b. 0.750 in.
    c. 0.500 in.
15. 0.3120 in.
17. 38.71 cm^2
19. $^{15}\!/_{16}$ in.

21. 4.796
23. 2.714 in.
25. 10.49 mm
27. 200°
29. ∠ A = 108°
    ∠ C = 72°
    ∠ E = 72°

UNIT 33 USE OF GRAPHS

1. a. Labor
   b. Supplies
   c. 17%

3. a. 1992
   b. 1988
   c. 1991
   d. 1,250

5. a. 900 hours
   b. October
   c. April
   d. October, November, and December

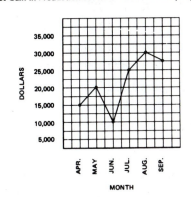

Net Gain in Production at Acme Wholesale Company

SECTION 9 APPLIED GEOMETRY

UNIT 34 LINES, SHAPES AND GEOMETRIC CONSTRUCTION

One method is given for the constructions. There are other possible solutions.

1. Obtuse
3. 90°
5. A = Cylinder
    B = Cone
    C = Isosceles triangle
    D = Equilateral (equi-
        angular) triangle
    E = Right triangle

F = Right-square
    pyramid
G = Eccentric circles
H = Triangular prism
I = Trapezoid
7. see Instructor's Guide
9. see Instructor's Guide
11. see Instructor's Guide

13. see Instructor's Guide
15. see Instructor's Guide
17. see Instructor's Guide
19. see Instructor's Guide
21. see Instructor's Guide
23. see Instructor's Guide
25. see Instructor's Guide
27. see Instructor's Guide

SECTION 10 APPLIED TRIGONOMETRY

UNIT 35 RIGHT TRIANGLES

1. Cosine 53° is the
    cofunction.
3. Sine 12° is the cofunction.
5. $\angle x = 77°\ 20'$
7. $\cos 27°\ 13' = 0.889281$
9. X = 2.0 inches
11. x = 64.344 inches
13. x = 28 inches

15. B = 6 in.
17. $\angle X = 29°\ 45'$
19. X = 248.42 mm
21. X = 3.362 inches
23. X = 4.424 inches
25. X = 0.323 inches
27. L = 3.858 inches

29. 3.54 in.
31. A = 3.83 in.
     C = 1.77 in.
33. 4.502 in.
35. A = 4.89 in.
     B = 3.70 in.
     C = 3.91 in.

UNIT 36 OBLIQUE TRIANGLES

1. 0.89662
3. $\cos 54°\ 33'$
5. −0.77522
7. −0.83867
9. a = 1.99 inches
    b = 1.30 inches

11. 3.77 sq. in.
13. $\angle A = 65°\ 8'$
     $\angle B = 93°\ 35'$
     $\angle C = 21°\ 17'$
15. 0.382 inches
17. The obtuse angle X is 112°

19. $\angle A = 79°\ 58'$
21. A = 3.536"
     B = 4.106"
     C = 2.79"
23. 3.69"